T0199235

Discrete-Time Recurrent Neural Control

Analysis and Applications

Automation and Control Engineering
Series Editors - Frank L. Lewis, Shuzhi Sam Ge, and Stjepan Bogdan

Discrete-Time Recurrent Neural Control

Analysis and Applications

Edgar N. Sánchez

CRC Press
Taylor & Francis Group
Boca Raton London New York

CRC Press is an imprint of the
Taylor & Francis Group, an **informa** business

CRC Press
Taylor & Francis Group
6000 Broken Sound Parkway NW, Suite 300
Boca Raton, FL 33487-2742

First issued in paperback 2022

© 2019 by Taylor & Francis Group, LLC
CRC Press is an imprint of Taylor & Francis Group, an Informa business

No claim to original U.S. Government works

ISBN-13: 978-1-138-55020-9 (hbk)
ISBN-13: 978-1-03-233896-5 (pbk)
DOI: 10.1201/9781315147185

**Visit the Taylor & Francis Web site at
http://www.taylorandfrancis.com**

**and the CRC Press Web site at
http://www.crcpress.com**

Dedication

To my wife: María de Lourdes,
our sons: Zulia Mayari, Ana María and Edgar Camilo,
and our grandsons: Esteban, Santiago and Amelia

Contents

SECTION I Analyses

SECTION II   Real-Time Applications

Preface

Neural networks are nowadays a well-established methodology for identification and control of general nonlinear and complex systems. Applying neural networks, control schemes can be developed to be robust in the presence of disturbances, parameter variations and modeling errors. The most utilized NN structures are feedforward networks and recurrent ones; the latter type is better suited to model and control of nonlinear systems. There exist different procedures to train neural networks, which normally face technical difficulties such as local minima, slow learning, and high sensitivity, among others. As a feasible alternative, methods based on Kalman filtering have been proposed.

There already exist results for trajectory tracking using neural networks; however most of them consider continuous-time systems. On the other hand, while a large number of publications are related to linear discrete-time control systems, the nonlinear case has not been considered to the same extent, even if discrete-time controllers are better fitted for real-time implementation.

Considering the above facts, the present book develops two discrete-time neural control schemes for trajectory tracking of nonlinear systems. Both of them are based on determining a model for the unknown system using a recurrent neural network, trained on-line with a Kalman filter, for identification. Once this model is obtained, two control methodologies are applied. First, block controllable forms combined with sliding modes are used, and then the inverse optimal control approach is employed. The scheme made up by the neural identifier and the control law constitutes an adaptive controller.

The book is organized in two sections. The first one covers accurate analysis of the properties of the proposed schemes, which is mainly done by means of the Lyapunov methodology; the last section presents real-time implementations of these schemes, which are performed using three-phases induction motors, widely used for industrial applications, and doubly fed induction generators, crucial for wind energy.

The book describes research results obtained during the last fourteen years at the Automatic Control Systems Laboratory of the CINVESTAV-IPN, from its name in Spanish: Center for Research and Advanced Studios of the National Polytechnic Institute, Guadalajara Campus.

Guadalajara, Mexico

December 2017

Acknowledgments

The author thanks *National Council for Science and Technology* (CONACyT for its name in Spanish, Consejo Nacional de Ciencia y Tecnología), Mexico, for financial support on Project Nos. 57801, 131678 and 257200, which allowed us to develop the research reported in this book.

He also thanks CINVESTAV-IPN (Center for Research and Advanced Studies of the National Polytechnic Institute), Mexico for providing him with the outstanding environment and resources to develop his research, from 1997 to date.

Additionally, he expresses his gratitude to Ronald G. Harley, professor at the School of Electrical and Computer Engineering of the Georgia Institute of Technology, Atlanta, Georgia, USA, for his support over the years and for giving permission to use the equipment at the Intelligent Power Infrastructure Consortium Laboratory of such institute.

The author is very grateful to his former PhD students, Alma Y. Alanis, professor at University of Guadalajara, Guadalajara, Mexico, Fernando Ornelas-Tellez, professor at Michoacan University, Morelia, Mexico and Riemann Ruiz-Cruz, professor at ITESO University, Guadalajara, Mexico, as well as to his present PhD students, Eduardo Quintero-Manriquez, María E. Antonio-Toledo, and Carlos J. Vega-Pérez; all of them, with their creativity, commitment and hard work, made this book possible.

Author

Edgar N. Sanchez was born in 1949 in Sardinata, Colombia, South America. He obtained a BSEE, with a major in power systems, from Universidad Industrial de Santander (UIS), Bucaramanga, Colombia in 1971, an MSEE from CINVESTAV-IPN (Center for Research and Advanced Studies of the National Polytechnic Institute), with a major in automatic control, Mexico City, Mexico, in 1974 and the Docteur Ingenieur degree in automatic control from Institut Nationale Polytechnique de Grenoble, France in 1980. He was granted a National Research Council Award as a research associate at NASA Langley Research Center, Hampton, Virginia, USA (January 1985 to March 1987). His research interests center on neural networks and fuzzy logic as applied to automatic control systems. Since January 1997, he has been with CINVESTAV-IPN, Guadalajara Campus, Mexico, as a professor of electrical engineering graduate programs. He has been the advisor of 24 PhD thesis and 42 MSc thesis students. He is also member of the Mexican National Research System (promoted to the highest rank, III, in 2005), the Mexican Academy of Science and the Mexican Academy of Engineering. He has published 7 books, more than 300 technical papers in international journals and conferences, and has served as associate editor and reviewer for different international journals and conferences. He has also been a member of many international IEEE and IFAC conference IPCs.

Section I

Analyses

1 Introduction

1.1 PRELIMINARIES

The ultimate goal of control engineering is to implement an automatic system which could operate with increasing independence from human actions in an unstructured and uncertain environment. Such a system can be called autonomous or intelligent. It would need only to be presented with a goal and would achieve its objective by learning through continuous interaction with its environment through feedback about its behavior [17].

One class of models which has the capability to implement this learning is the artificial neural networks. Indeed, the neural morphology of the nervous system is quite complex to analyze. Nevertheless, simplified analogies have been developed, which could be used for engineering applications. Based on these simplified understandings, artificial neural networks are built [9].

An artificial neural network is a massively parallel distributed processor, inspired by biological neural networks, which can store experimental knowledge and have it available for use. An artificial neural network consists of a finite number of neurons (structural element), which are interconnected to each other. It has some similarities with the brain: knowledge is acquired through a learning process, and interneuron connectivity called synaptic weights are used to store this knowledge, among others [17].

The research on neural networks, since its rebirth in the early 1980s, promotes great interest principally due to the capability of static neural networks to approximate arbitrarily well any continuous function. Besides, in recent years, the use of recurrent neural networks has increased; their information processing is described by differential equations for continuous time or by difference equations for discrete time [9].

Using neural networks, control algorithms can be developed to be robust to uncertainties and modeling errors. The most used neural network structures are: *Feedforward* networks and *Recurrent* ones [1, 18]. The last type offers a tool that is better suited to model and control nonlinear systems [15].

There exist different training algorithms for neural networks, which, however, normally encounter technical problems such as local minima, slow learning, and high sensitivity to initial conditions, among others. As a viable alternative, new training algorithms, e.g., those based on Kalman filtering, have been proposed [8, 9, 19]. Due to the fact that training a neural network typically results in a nonlinear problem, the Extended Kalman Filter (EKF) is a common tool to use, instead of a linear Kalman filter [9].

There already exist publications about trajectory tracking using neural networks ([4], [10], [11], [12], [13], [15], [16], [18]); in most of them, the design methodology is based on the Lyapunov approach. However, the majority of those works were developed for continuous-time systems. On the other hand, while extensive literature is available for linear discrete-time control systems, nonlinear discrete-time control design techniques have not been discussed to the same degree. For nonlinear discrete-time systems, the control problem is more complex due to couplings among subsystems, inputs and outputs [2, 7, 11]. Besides, discrete-time neural networks are better fitted for real-time implementations. There are two advantages to working in a discrete-time framework: a) appropriate technology can be used to implement digital controllers rather than analog ones b) the synthesized controller is directly implemented in a digital processor. Therefore, control methodologies developed for discrete-time nonlinear systems can be implemented in real systems more effectively. In this book, it is considered a class of nonlinear systems, the affine nonlinear one, which represents a great variety of them, most of which are approximate discretizations of continuous-time systems.

This book presents two types of controllers for trajectory tracking of unknown discrete-time nonlinear systems with external disturbances and internal uncertainties based on two approaches; the first one is based on the sliding mode technique, and the second one uses inverse optimal control; both of them are designed based on a neural model, and the applicability of the proposed controllers is illustrated, via simulations and real-time results. As a special case, the proposed control scheme is applied to electric machines. It is worth mentioning that if a continuous-time control scheme is real-time implemented, there is no guarantee that it preserves its properties, such as stability margins and adequate performance. Even worse, it is known that continuous time schemes could become unstable after sampling.

To control a system is to force it to behave in a desired way. How to express this "desired behavior" depends primarily on the task to be solved; however, the dynamics of the system, the actuators, the measurement equipment, the available computational power, etc., influence the formulation of the desired behavior as well. Although the desired behavior obviously is very dependent of the application, the need to express it in mathematical terms suited for practical design of control systems seriously limits the means of expression. At the higher level, it is customary to distinguish two basic types of problems [14]:

Regulation. The fundamental desired behavior is to keep the output of the system at a constant level regardless of the disturbances acting on the system.

Tracking. The fundamental desired behavior is to force the system output to track a reference trajectory closely.

Neural networks (NNs) have become a well-established methodology as exemplified by their applications to identification and control of general nonlinear and complex systems [6, 14]; the use of high order neural networks for modeling and learning has recently increased [18]. Specifically, the problem of designing robust neural controllers for nonlinear systems with parametric uncertainties, unmodeled dynamics and external disturbances, which guarantees stability and trajectory tracking, has received increasing attention lately. Using neural networks, control algorithms can be developed to be robust in the presence of such events.

Neural controller synthesis can be approached in two different ways:

Direct control system design. "Direct" means that the controller is a neural network. A neural network controller is often advantageous when the real-time platform available prohibits complicated solutions. The implementation is simple while the design and tuning are difficult. With a few exceptions this class of designs is model-based in the sense that a model of the system is required in order to determine the controller.

Indirect control system design. This class of designs is always model-based. The idea is to use a neural network to model the system to be controlled, which is then employed in a more "conventional" controller design. The model is typically trained in advance, but the controller is designed on-line. As yo will see, the indirect design is very flexible; thus it is the most appropriate.

The increasing use of NNs for modeling and control of nonlinear systems is in great part due to the following features, which make them particularly attractive [5]:

- NNs are universal approximators. It has been proven that any continuous nonlinear function can be approximated arbitrarily well over a compact set by a multilayer neural network which consists of one or more hidden layers [3].

- Learning and adaptation. The intelligence of neural networks comes from their generalization ability with respect to unknown data. On-line adaptation of the weights is also possible.

1.2 MOTIVATION

Taking into account the facts exposed above, the need to synthesize control algorithms for Multiple Input Multiple Output (MIMO) discrete-time nonlinear systems based on neural networks is obvious. These algorithms should be robust to external disturbances as well as parametric variations.

On the other hand, in most nonlinear control designs, it is usually assumed that the system model is previously known, as well as its parameters and disturbances. In practice, however, only part of this model is known. For this reason, identification remains an important topic, particularly neural identification.

Therefore, the major motivation for this book is to develop alternative methodologies, which allow the design of robust controllers for discrete-time nonlinear systems with unknown dynamics.

Finally, there only exist a few published results on real-time implementations of neural controllers; so this book contains mathematical analysis, simulation examples and real-time implementation for all the proposed schemes.

1.3 OBJECTIVES

The main objectives of this book are stated as follows:

- To synthesize a neural identifier for a class of MIMO discrete-time nonlinear systems, using a training algorithm based on an EKF.

- To synthesize a scheme for output trajectory tracking based on a Recurrent High Order Neural Network (RHONN) trained with an EKF, to identify a MIMO discrete-time nonlinear system and based on the neural model, to design a control law by the block control and sliding mode techniques.

- To synthesize a scheme for output trajectory tracking based on a Recurrent High Order Neural Network (RHONN) trained with an EKF, to identify a MIMO discrete-time nonlinear system, and based on the neural model, to define a control law by the inverse optimal control technique.

- To establish the stability analyses, using the Lyapunov approach, for each one of the proposed schemes.

- To implement real-time experiments for each one of the proposed schemes.

1.4 BOOK STRUCTURE

This book presents a solution for the trajectory tracking of unknown nonlinear systems based on two schemes. For the first one, an indirect method, is solved with the block control and the sliding mode techniques, under the assumption of complete access to the state; the second one considers an indirect method, solved with the inverse optimal control technique, under the same assumption. Both schemes are developed in discrete-time.

This book is organized as follows.

In *Chapter 2*, mathematical preliminaries are introduced, including stability definitions, artificial neural network foundations, and the principle of separation for discrete-time nonlinear systems.

Then in *Chapter 3*, the identified model is used to design a block control form controller, based on sliding mode. The training of the neural networks is performed on-line using an extended Kalman filter.

After that, in *Chapter 4*, the identified model is used to design an inverse optimal neural control. The training of the neural networks is performed on-line using an extended Kalman filter.

Chapter 5 includes real-time results for the neural identifier and the two control schemes developed in the previous chapters are applied to a three-phase induction motor.

In *Chapter 6*, real-time results are presented for the neural identifier and the controllers previously explained are applied to a double feed induction generator (DFIG).

Chapter 7 presents relevant conclusions.

Additionally, an appendix is included at the end of this book. In this appendix, the DFIG and DC Link mathematical model development is detailed.

1.5 NOTATION

Through this book, we use the following notation:

$k \in 0 \cup \mathbb{Z}^+$	Sampling step
$|\bullet|$	Absolute value
$\|\bullet\|$	Euclidian norm for vectors and any adequate norm for matrices
$S(\bullet)$	Sigmoid function
$x \in \Re^n$	Plant state
$\hat{x} \in \Re^n$	Neural network state
$w_i \in \Re^L$	i-th neural network estimated weight vector
$w_i^* \in \Re^L$	i-th neural network ideal weight vector

$L_i \in \Re$ Number of high-order connections

$u \in \Re^m$ Control action

$u^* \in \Re^m$ Ideal control action

$\rho \in \Re^m$ Neural network external input

$z_i \in \Re^{L_i}$ High-order terms

$K \in \Re^{L_i \times m}$ Kalman gain matrix

$P \in \Re^{L_i \times L_i}$ Associated prediction error covariance matrix

$Q \in \Re^{L_i \times L_i}$ Associated state noise covariance matrix

$R \in \Re^{m \times m}$ Associated measurement noise covariance matrix

$g_i \in \Re$ i-th neural observer gain

$r \in \Re$ Number of blocks

$n_i \in \Re$ Dimension of the i-th block

$S_D \in \Re^{n_r}$ Sliding manifold

$\mathbf{k}_i \in \Re$ Control gain of the i-th block

$\mathbf{z}_i \in \Re^{n_i}$ State transformation of the i-th block

$e \in \Re^p$ Output error

$\widetilde{x} \in \Re^n$ State observer error

$\widetilde{w}_i \in \Re^{L_i}$ Weights estimation error

1.6 ACRONYMS

BIBS   Bounded-Input Bounded-State

CLA   Control Accelerator

CLF   Control Lyapunov Function

DARE   Discrete-Time Algebraic Riccati Equation

DC   Direct Current

DFIG   Doubly Fed Induction Generator

DT   Discrete-Time

EKF   Extended Kalman Filter

FOC   Field Oriented Control

GAS	Globally Asymptotically Stable
GS	Globally Stable
GSC	Grid Side Converter
HJB	Hamilton-Jacobi-Bellman
HJI	Hamilton-Jacobi-Isaacs
IOC	Inverse Optimal Control
ISS	Input-to-State Stable
KF	Kalman Filtering
LQR	Linear Quadratic Regulator
NIOC	Neural Inverse Optimal Control
NN	Neural Network
PWM	Pulse-Width Modulation
QEP	Quadrature Encoder Pulse
RCP	Rapid Control Prototyping
RHONN	Recurrent High Order Neural Network
RHS	Right-Hand Side
RNN	Recurrent Neural Network
RSC	Rotor Side Converter
SCI	Serial Communications Interface
SMC	Sliding Mode Control
SG	Speed-Gradient
SG-IOC	Speed-Gradient Inverse Optimal Control
SG-IONC	Speed-Gradient Inverse Optimal Neural Control
SGUUB	Semiglobally Uniformly Ultimately Bounded
SVM	Space Vector Modulation

REFERENCES

1. A. Y. Alanis, *Neural network training using Kalman Filtering*, Master's Dissertation, Cinvestav, Unidad Guadalajara, Guadalajara Jalisco Mexico, 2004 (in spanish).

2. F. Chen and H. Khalil, Adaptive control of a class of nonlinear discrete-time systems using neural networks, *IEEE Transactions on Automatic Control*, vol. 40, no. 5, pp. 791–801, 1995.

3. N. Cotter. The Stone-Weiertrass theorem and its application to neural networks, *IEEE Transactions on Neural Networks*, vol. 1, no. 4, pp. 290-295, 1990.

4. L. A. Feldkamp, D. V. Prokhorov and T. M. Feldkamp, Simple and conditioned adaptive behavior from Kalman filter trained recurrent networks, *Neural Networks*, vol. 16, pp. 683–689, 2003.

5. R. A. Felix, *Variable Structure Neural Control*, PhD Dissertation, Cinvestav, Unidad Guadalajara, Guadalajara Jalisco Mexico, 2004.

6. S.S. Ge, T.H. Lee, and C.J. Harris, Adaptive Neural Network Control for Robotic Manipulators, World Scientific, Singapore, 1998.

7. S. S. Ge, J. Zhang and T. H. Lee, Adaptive neural network control for a class of MIMO nonlinear systems with disturbances in discrete-time, *IEEE Transactions on Systems, Man and Cybernetics*, Part B, vol. 34, no. 4, August, 2004.

8. R. Grover and P. Y. C. Hwang, *Introduction to Random Signals and Applied Kalman Filtering*, 2nd ed., John Wiley and Sons, N. Y., USA, 1992.

9. S. Haykin, *Kalman Filtering and Neural Networks*, John Wiley and Sons, N. Y., USA, 2001.

10. S. Jagannathan, Control of a class of nonlinear discrete-time systems using multilayer neural networks, *IEEE Transactions on Neural Networks*, vol. 12, no. 5, pp. 1113–1120, 2001.

11. F.L. Lewis, J. Campos, and R. Selmic, *Neuro-Fuzzy Control of Industrial Systems with Actuator Nonlinearities*, Society of Industrial and Applied Mathematics Press, Philadelphia, 2002.

12. F.L. Lewis, S. Jagannathan, and A. Yesildirek, "*Neural Network Control of Robot Manipulators and Nonlinear Systems*", Taylor and Francis, London, 1999.

13. K.S. Narendra and K. Parthasarathy, "Identification and control of dynamical systems using neural networks," *IEEE Transactions on. Neural Networks*, vol. 1, pp. 4–27, Mar. 1990.

14. M. Norgaard, O. Ravn, N. K. Poulsen and L. K. Hansen, *Neural Networks for Modelling and Control of Dynamic Systems*, Springer Verlag, New York, USA, 2000.

15. A. S. Poznyak, E. N. Sanchez and W. Yu, *Differential Neural Networks for Robust Nonlinear Control*, World Scientific, Singapore, 2001.

16. G.A. Rovithakis and M.A. Chistodoulou, *Adaptive Control with Recurrent High-Order Neural Networks*, Springer Verlag, Berlin, Germany, 2000.

17. E. N. Sanchez, A. Y. Alanis and G. Chen, Recurrent neural networks trained with Kalman filtering for discrete chaos reconstruction, *Dynamics of Continuous, Discrete and Impulsive Systems Series B (DCDIS_B)*, vol. 13, pp. 1–18, 2006.

18. E. N. Sanchez and L. J. Ricalde, Trajectory tracking via adaptive recurrent neural control with input saturation, *Proceedings of International Joint Conference on Neural Networks'03*, Portland, Oregon, USA, July, 2003.

19. S. Singhal and L. Wu, Training multilayer perceptrons with the extended Kalman algorithm, in D. S. Touretzky (ed.), *Advances in Neural Information Processing Systems*, Vol. 1, pp. 133–140, Morgan Kaufmann, San Mateo, CA, USA, 1989.

2 Mathematical Preliminaries

This chapter briefly describes useful results on optimal control theory, Lyapunov stability, passivity, and neural identification, required in future chapters.

2.1 OPTIMAL CONTROL

This section briefly discusses the optimal control methodology and its limitations.

Consider the affine-in-the-input discrete-time nonlinear system:

$$x_{k+1} = f(x_k) + g(x_k) u_k, \qquad x_0 = x(0), \qquad (2.1)$$

where $x_k \in \mathbb{R}^n$ is the state of the system at time $k \in \mathbb{Z}^+ \cup 0 = \{0, 1, 2, \ldots\}$, $u_k \in \mathbb{R}^m$ is the input, $f : \mathbb{R}^n \to \mathbb{R}^n$ and $g : \mathbb{R}^n \to \mathbb{R}^{n \times m}$ are smooth mappings, $f(0) = 0$ and $g(x_k) \neq 0$ for all $x_k \neq 0$.

For system (2.1), it is desired to determine a control law $u_k = \bar{u}(x_k)$ which minimizes the following cost functional:

$$V(x_k) = \sum_{n=k}^{\infty} \left(l(x_n) + u_n^T R u_n \right), \qquad (2.2)$$

where $V : \mathbb{R}^n \to \mathbb{R}^+$ is a performance measure [13]; $l : \mathbb{R}^n \to \mathbb{R}^+$ is a positive semidefinite[1] function weighting the performance of the state vector x_k, and $R : \mathbb{R}^n \to \mathbb{R}^{m \times m}$ is a real symmetric and positive definite[2] matrix weighting the control effort expenditure. The entries of R could be functions of the system state in order to vary the weighting on control efforts according to the state value [13].

---

[1] A function $l(z)$ is a positive semidefinite (or nonnegative definite) function if for all vectors z, $l(z) \geq 0$. In other words, there are vectors z for which $l(z) = 0$, and for all others z, $l(z) > 0$ [13].

[2] A real symmetric matrix R is positive definite if $z^T R z > 0$ for all $z \neq 0$ [13].

Equation (2.2) can be rewritten as

$$
\begin{aligned}
V(x_k) &= l(x_k) + u_k^T R u_k + \sum_{n=k+1}^{\infty} l(x_n) + u_n^T R u_n \\
&= l(x_k) + u_k^T R u_k + V(x_{k+1}).
\end{aligned}
\tag{2.3}
$$

From Bellman's optimality principle [17], it is known that, for the infinite horizon optimization case, the value function $V^*(x_k)$ becomes time invariant and satisfies the discrete-time Bellman equation [17]

$$
V^*(x_k) = \min_{u_k} \left\{ l(x_k) + u_k^T R u_k + V^*(x_{k+1}) \right\}.
\tag{2.4}
$$

Note that the Bellman equation is solved backwards in time [17].

In order to establish the conditions that the optimal control law must satisfy, we define the discrete-time Hamiltonian $\mathcal{H}(x_k, u_k)$ as

$$
\mathcal{H}(x_k, u_k) = l(x_k) + u_k^T R u_k + V^*(x_{k+1}) - V^*(x_k),
\tag{2.5}
$$

which is used to obtain the control law u_k by calculating

$$
\min_{u_k} \mathcal{H}(x_k, u_k).
$$

The value of u_k, which achieves this minimization, is a feedback control law denoted as $u_k = \bar{u}(x_k)$, then

$$
\min_{u_k} \mathcal{H}(x_k, u_k) = \mathcal{H}(x_k, \bar{u}(x_k)).
$$

A necessary condition, which this feedback optimal control law $\bar{u}(x_k)$ must satisfy [13], is

$$
\mathcal{H}(x_k, \bar{u}(x_k)) = 0.
\tag{2.6}
$$

$\bar{u}(x_k)$ is obtained by calculating the gradient of the Right-Hand Side (RHS) of (2.5) with respect to u_k [17]

$$
\begin{aligned}
0 &= 2Ru_k + \frac{\partial V^*(x_{k+1})}{\partial u_k} \\
&= 2Ru_k + g^T(x_k)\frac{\partial V^*(x_{k+1})}{\partial x_{k+1}}.
\end{aligned}
\tag{2.7}
$$

Therefore, the optimal control law is formulated as

$$
\begin{aligned}
u_k^* &= \bar{u}(x_k) \\
&= -\frac{1}{2}R^{-1}g^T(x_k)\frac{\partial V^*(x_{k+1})}{\partial x_{k+1}},
\end{aligned}
\tag{2.8}
$$

which is a state feedback control law $\bar{u}(x_k)$ with $\bar{u}(0) = 0$. Hence, the boundary condition $V(0) = 0$ in (2.2) and (2.3) is satisfied for $V(x_k)$, and V becomes a Lyapunov function; u_k^* is used to emphasize that u_k is optimal.

Moreover, if $\mathcal{H}(x_k, u_k)$ is a quadratic form in u_k and $R > 0$, then

$$
\frac{\partial^2 \mathcal{H}(x_k, u_k)}{\partial u_k^2} > 0
$$

holds as a sufficient condition such that optimal control law (2.8) (globally [13]) minimizes $\mathcal{H}(x_k, u_k)$ and the performance index (2.2) [17]. Substituting (2.8) into (2.4), we obtain

$$
\begin{aligned}
V^*(x_k) &= l(x_k) + \left(-\frac{1}{2}R^{-1}g^T(x_k)\frac{\partial V^*(x_{k+1})}{\partial x_{k+1}}\right)^T \\
&\quad \times R\left(-\frac{1}{2}R^{-1}g^T(x_k)\frac{\partial V^*(x_{k+1})}{\partial x_{k+1}}\right) + V^*(x_{k+1}) \\
&= l(x_k) + V^*(x_{k+1}) + \frac{1}{4}\frac{\partial V^{*T}(x_{k+1})}{\partial x_{k+1}}g(x_k)R^{-1}g^T(x_k)\frac{\partial V^*(x_{k+1})}{\partial x_{k+1}}
\end{aligned}
\tag{2.9}
$$

which can be rewritten as

$$
l(x_k) + V^*(x_{k+1}) - V^*(x_k) + \frac{1}{4}\frac{\partial V^{*T}(x_{k+1})}{\partial x_{k+1}}g(x_k)R^{-1}g^T(x_k)\frac{\partial V^*(x_{k+1})}{\partial x_{k+1}} = 0. \tag{2.10}
$$

Equation (2.10) is known as the discrete-time HJB equation [17]. Solving this partial-differential equation for $V^*(x_k)$ is not straightforward. This is one of the main draw-backs in discrete-time optimal control for nonlinear systems. To overcome this problem, we propose using inverse optimal control.

2.2 LYAPUNOV STABILITY

In order to establish stability, we recall important related properties.

Definition 2.1: Radially Unbounded Function [12] A positive definite function $V(x_k)$ satisfying $V(x_k) \to \infty$ as $\|x_k\| \to \infty$ is said to be radially unbounded.

Definition 2.2: Decrescent Function [12] A function $V : \mathbb{R}^n \to \mathbb{R}$ is said to be decrescent if there is a positive definite function β such that the following inequality holds:

$$V(x_k) \le \beta(\|x_k\|), \qquad \forall k \ge 0.$$

Definition 2.3: $\mathcal{K}$ and $\mathcal{K}_\infty$ functions [21] A function $\gamma : \mathbb{R}_{\ge 0} \to \mathbb{R}_{\ge 0}$ is a $\mathcal{K}$-function if it is continuous, strictly increasing and $\gamma(0) = 0$; it is a $\mathcal{K}_\infty$-function if it is a $\mathcal{K}$-function and also $\gamma(s) \to \infty$ as $s \to \infty$; and it is a positive definite function if $\gamma(s) > 0$ for all $s > 0$, and $\gamma(0) = 0$.

Definition 2.4: $\mathcal{K}\mathcal{L}$-function [21] A function $\beta : \mathbb{R}_{\ge 0} \times \mathbb{R}_{\ge 0} \to \mathbb{R}_{\ge 0}$ is a $\mathcal{K}\mathcal{L}$-function if, for each fixed $t \ge 0$, the function $\beta(\cdot, t)$ is a $\mathcal{K}$-function, and for each fixed $s \ge 0$, the function $\beta(s, \cdot)$ is decreasing and $\beta(s, t) \to 0$ as $t \to \infty$. $\mathbb{R}_{\ge 0}$ means nonnegative real numbers.

Theorem 2.1: Global Asymptotic Stability [15]

The equilibrium point $x_k = 0$ of (2.1) is globally asymptotically stable if there exists a function $V : \mathbb{R}^n \to \mathbb{R}$ such that (i) V is a positive definite function, decrescent and radially unbounded, and (ii) $-\Delta V(x_k, u_k)$ is a positive definite function, where $\Delta V(x_k, u_k) = V(x_{k+1}) - V(x_k)$. ■

Theorem 2.2: Exponential Stability [27]

Suppose that there exists a positive definite function $V : \mathbb{R}^n \to \mathbb{R}$ and constants c_1, c_2, $c_3 > 0$ and $p > 1$ such that

$$c_1 \|x\|^p \leq V(x_k) \leq c_2 \|x\|^p \qquad (2.11)$$

$$\Delta V(x_k) \leq -c_3 \|x\|^p, \quad \forall k \geq 0, \quad \forall x \in \mathbb{R}^n. \qquad (2.12)$$

Then $x_k = 0$ is an exponentially stable equilibrium for system (2.1). ■

Clearly, exponential stability implies asymptotic stability. The converse is, however, not true.

Due to the fact that the inverse optimal control is based on a Lyapunov function, we establish the following definitions.

Definition 2.5: Control Lyapunov Function [2, 11] Let $V(x_k)$ be a radially unbounded function, with $V(x_k) > 0$, $\forall x_k \neq 0$ and $V(0) = 0$. If for any $x_k \in \mathbb{R}^n$ there exist real values u_k such that

$$\Delta V(x_k, u_k) < 0,$$

where we define the Lyapunov difference as $\Delta V(x_k, u_k) = V(f(x_k) + g(x_k) u_k) - V(x_k)$, then $V(\cdot)$ is said to be a discrete-time control Lyapunov function (CLF) for system (2.1).

Assumption 2.1 Let us assume that $x = 0$ is an equilibrium point for (2.1), and that there exists a control Lyapunov function $V(x_k)$ such that

$$\alpha_1(\|x_k\|) \leq V(x_k) \leq \alpha_2(\|x_k\|) \tag{2.13}$$

$$\Delta V(x_k, u_k) \leq -\alpha_3(\|x_k\|), \tag{2.14}$$

where α_1, α_2, and α_3 are class $\mathscr{K}_\infty$ functions[3] and $\|\cdot\|$ denotes the usual Euclidean norm. Then, the origin of the system is an asymptotically stable equilibrium point by means of u_k as input.

The existence of this CLF is guaranteed by the converse theorem of the Lyapunov stability theory [3].

As a special case, the calculus of class $\mathscr{K}_\infty$– functions in (2.13) simplifies when they take the special form $\alpha_i(r) = \kappa_i r^c$, $\kappa_i > 0$, $c = 2$, and $i = 1, 2$. In particular, for a quadratic positive definite function $V(x_k) = \frac{1}{2} x_k^T P x_k$, with P a positive definite and symmetric matrix, inequality (2.13) results in

$$\lambda_{min}(P)\|x\|^2 \leq x_k^T P x_k \leq \lambda_{max}(P)\|x\|^2, \tag{2.15}$$

where $\lambda_{min}(P)$ is the minimum eigenvalue of matrix P and $\lambda_{max}(P)$ is the maximum eigenvalue of matrix P.

2.3 ROBUST STABILITY ANALYSIS

This section reviews stability results for disturbed nonlinear systems, for which non-vanishing disturbances are considered. We can no longer study the stability of the

---

[3] α_i, $i = 1, 2, 3$ belong to class $\mathscr{K}_\infty$ functions because later we will select a radially unbounded function $V(x_k)$.

origin as an equilibrium point, nor should we expect the solution of the disturbed system to approach the origin as $k \to \infty$. The best we can hope for is that if the disturbance is small in some sense, then the system solution will be ultimately bounded by a small bound [12], which connects to the concept of ultimate boundedness.

Definition 2.6: Ultimate Bound [6, 12] The solutions of (2.1) with $u_k = 0$ are said to be uniformly ultimately bounded if there exist positive constants b and c, and for every $a \in (0,c)$ there is a positive constant $T = T(a)$, such that

$$\|x_0\| < a \Rightarrow \|x_k\| \leq b, \quad \forall k \geq k_0 + T, \tag{2.16}$$

where k_0 is the initial time instant. They are said to be globally uniformly ultimately bounded if (2.16) holds for arbitrarily large a. The constant b in (2.16) is known as the *ultimate bound*.

Definition 2.7: BIBS [21] System (2.1) is uniformly bounded-input bounded-state (BIBS) stable with respect to u_k, if bounded initial states and inputs produce uniformly bounded trajectories.

Definition 2.8: ISS Property [19, 21] System (2.1) is (globally) input-to-state stable (ISS) with respect to u_k if there exist a $\mathcal{KL}$– function β and a $\mathcal{K}$– function γ such that, for each input $u \in l_\infty^m$ and each $x_0 \in \mathbb{R}^n$, it holds that the solution of (2.1) satisfies

$$\|x_k\| \leq \beta(\|x_0\|, k) + \gamma\left(\sup_{\tau \in [k_0, \infty)} \|u_\tau\| \right), \tag{2.17}$$

where $\sup_{\tau \in [k_0, \infty)} \{\|u_\tau\| : \tau \in \mathbb{Z}^+\} < \infty$, which is denoted by $u \in \ell_\infty^m$.

Thus, system (2.1) is said to be ISS if property (2.17) is satisfied [14]. The interpre-

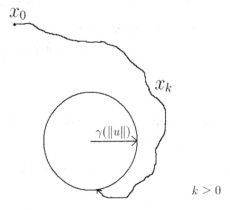

FIGURE 2.1 System trajectories with the ISS property.

tation of (2.17) is the following: for a bounded input u, the system solution remains in the ball of radius $\beta(\|x_0\|,k) + \gamma\left(\sup_{\tau\in[k_0,\infty)}\|u_\tau\|\right)$. Furthermore, as k increases, all trajectories approach the ball of radius $\gamma\left(\sup_{\tau\in[k_0,\infty)}\|u_\tau\|\right)$, i.e., all trajectories will be ultimately bounded with ultimate bound γ. Due to the fact that γ is of class $\mathscr{K}$, this ball is a small neighborhood of the origin whenever $\|u\|$ is small (see Figure 2.1). ISS is used to analyze stability of the solutions for disturbed nonlinear systems. The ISS property captures the notion of BIBS stability.

Definition 2.9: Asymptotic Gain Property [21] System (2.1) is said to have $\mathscr{K}-$ asymptotic gain if there exists some $\gamma\in\mathscr{K}$ such that

$$\lim_{k\to\infty}\|x_k(x_0,u)\| \leq \lim_{k\to\infty}\gamma(\|u_k\|),\qquad(2.18)$$

for all $x_0 \in \mathbb{R}^n$.

Theorem 2.3: ISS System [21]

Consider system (2.1). The following are equivalent:

1. It is ISS.

2. It is BIBS and it admits $\mathcal{K}$ – asymptotic gain.

∎

Let ℓ_d be the Lipschitz constant such that for all β_1 and β_2 in some bounded neighborhood of (x_k, u_k), the Lyapunov function $V(x_k)$ satisfies the condition [24]

$$\|V(\beta_1) - V(\beta_2)\| \le \ell_d \|\beta_1 - \beta_2\|, \quad \ell_d > 0. \tag{2.19}$$

Definition 2.10: ISS – Lyapunov Function [21] A continuous function V on $\mathbb{R}^n$ is called an ISS–Lyapunov function for system (2.1) if

$$\alpha_1(\|x_k\|) \le V(x_k) \le \alpha_2(\|x_k\|), \tag{2.20}$$

holds for some α_1, $\alpha_2 \in \mathcal{K}_\infty$, and

$$V(f(x_k, u_k)) - V(x_k) \le -\alpha_3(\|x_k\|) + \sigma(\|u_k\|), \tag{2.21}$$

for some $\alpha_3 \in \mathcal{K}_\infty$, $\sigma \in \mathcal{K}$. A smooth ISS–Lyapunov function is one which is smooth.

Note that if $V(x_k)$ is an ISS–Lyapunov function for (2.1), then $V(x_k)$ is a DT Lyapunov function for the 0-input system $x_{k+1} = f(x_k) + g(x_k)0$.

Proposition 2.1

If system (2.1) admits an ISS–Lyapunov function, then it is ISS [21].

Now, consider the disturbed system

$$x_{k+1} = f(x_k) + g(x_k) u_k + d_k, \qquad x_0 = x(0), \qquad (2.22)$$

where $x_k \in \mathbb{R}^n$ is the state of the system at time $k \in \mathbb{Z}^+$, $u_k \in \mathbb{R}^m$ is the control, $d_k \in \mathbb{R}^n$ is the disturbance term, $f: \mathbb{R}^n \to \mathbb{R}^n$ and $g: \mathbb{R}^n \to \mathbb{R}^{n \times m}$ are smooth mappings, $f(0) = 0$. d_k could result from modeling errors, aging, or uncertainties and disturbances which exist for any realistic problem [12].

Definition 2.11: ISS–CLF Function A smooth positive definite radially unbounded function $V: \mathbb{R}^n \to \mathbb{R}$ is said to be an ISS–CLF for system (2.22) if there exists a class $\mathcal{K}_\infty$ function ρ such that the following inequalities hold $\forall x \neq 0$ and $\forall d \in \mathbb{R}^n$:

$$\alpha_1(\|x_k\|) \leq V(x_k) \leq \alpha_2(\|x_k\|), \qquad (2.23)$$

for some α_1, $\alpha_2 \in \mathcal{K}_\infty$, and

$$\|x_k\| \geq \rho(\|d_k\|) \Rightarrow \inf_{u_k \in \mathbb{R}^m} \Delta V_d(x_k, d_k) < -\alpha_3(\|x_k\|), \qquad (2.24)$$

where $\Delta V_d(x_k, d_k) := V(x_{k+1}) - V(x_k)$ and $\alpha_3 \in \mathcal{K}_\infty$.

Comment 2.1 The connection between the existence of a Lyapunov function and the input-to-state stability is that an estimate of the gain function γ in (2.17) is $\gamma = \alpha_1^{-1} \circ \alpha_2 \circ \rho$, where $\circ$ means composition[4] of functions with α_1 and α_2 as defined in (2.23) [14].

Note that if $V(x_k)$ is an ISS–control Lyapunov function for (2.22), then $V(x_k)$ is a control Lyapunov function for the 0-disturbance system $x_{k+1} = f(x_k) + g(x_k) u_k$.

---

[4] $\alpha_1(\cdot) \circ \alpha_2(\cdot) = \alpha_1(\alpha_2(\cdot))$.

Proposition 2.2: ISS–CLF System

If system (2.22) admits an ISS–CLF, then it is ISS.

2.3.1 OPTIMAL CONTROL FOR DISTURBED SYSTEMS

For a disturbed discrete-time nonlinear system (2.22), the Bellman equation becomes the Isaacs equation described by

$$V(x_k) = \min_{u_k} \left\{ l(x_k) + u_k^T R(x_k) u_k + V(x_k, u_k, d_k) \right\}, \tag{2.25}$$

and the Hamilton–Jacobi–Isaacs (HJI) equation associated with system (2.22) and cost functional (2.2) is

$$
\begin{aligned}
0 &= \inf_u \sup_{d \in \mathscr{D}} \left\{ l(x_k) + u_k^T R(x_k) u_k + V(x_{k+1}) - V(x_k) \right\} \\
&= \inf_u \sup_{d \in \mathscr{D}} \left\{ l(x_k) + u_k^T R(x_k) u_k + V(x_k, u_k, d_k) - V(x_k) \right\}
\end{aligned} \tag{2.26}
$$

where $\mathscr{D}$ is the set of locally bounded functions, and function $V(x_k)$ is unknown. However, determining a solution of the HJI equation (2.26) for $V(x_k)$ with (2.8) is the main drawback of robust optimal control; this solution may not exist or may be pretty difficult to solve [8]. Note that $V(x_{k+1})$ in (2.26) is a function of the disturbance term d_k.

2.4. PASSIVITY

Let us consider a nonlinear affine system and an output given as

$$x_{k+1} = f(x_k) + g(x_k) u_k, \qquad x_0 = x(0) \tag{2.27}$$

$$y_k = h(x_k) + J(x_k) u_k \tag{2.28}$$

where $x_k \in \mathbb{R}^n$ is the state of the system at time k, output $y_k \in \mathbb{R}^m$; $h(x_k) : \mathbb{R}^n \to \mathbb{R}^m$, and $J(x_k) : \mathbb{R}^n \to \mathbb{R}^{m \times m}$ are smooth mappings. We assume $h(0) = 0$.

Definition 2.12: Passivity [4] System (2.27)–(2.28) is said to be passive if there exists a nonnegative function $V(x_k)$, called the storage function, such that for all u_k,

$$V(x_{k+1}) - V(x_k) \leq y_k^T u_k, \tag{2.29}$$

where $(\cdot)^T$ denotes transpose.

This storage function may be selected as a CLF candidate if it is a positive definite function [25]. It is worth noting that the output which renders the system passive is not in general the variable we wish to control, and it is used only for control synthesis.

Definition 2.13: Zero–State Observable System [5] A system (2.27)–(2.28) is locally zero-state observable (respectively, locally zero-state detectable) if there exists a neighborhood $\mathscr{Z}$ of $x_k = 0$ in $\mathbb{R}^n$ such that for all $x_0 \in \mathscr{Z}$

$$y_k|_{u_k=0} = h(\phi(k,x_0,0)) = 0 \quad \forall k \implies x_k = 0 \left(respectively \lim_{k \to \infty} \phi(k,x_0,0) = 0 \right),$$

where $\phi(k,x_0,0) = f^k(x_k)$ is the trajectory of the unforced dynamics $x_{k+1} = f(x_k)$ with initial condition x_0. If $\mathscr{Z} = \mathbb{R}^n$, the system is zero-state observable (respectively, zero-state detectable).

Additionally, the following definition is introduced.

Definition 2.14: Feedback Passive System System (2.27)–(2.28) is said to be feedback passive if there exists a passivation law

$$u_k = \alpha(x_k) + v_k, \qquad \alpha, v \in \mathbb{R}^m, \tag{2.30}$$

with a smooth function $\alpha(x_k)$ and a storage function $V(x)$, such that system (2.27) with (2.30), described by

$$x_{k+1} = \bar{f}(x_k) + g(x_k)v_k, \qquad x_0 = x(0), \tag{2.31}$$

and output

$$\bar{y}_k = \bar{h}(x_k) + J(x_k)v_k, \tag{2.32}$$

satisfies relation (2.29) with v_k as the new input, where $\bar{f}(x_k) = f(x_k) + g(x_k)\alpha(x_k)$ and $\bar{h} : \mathbb{R}^n \to \mathbb{R}^m$ is a smooth mapping, which will be defined later, with $\bar{h}(0) = 0$.

Roughly speaking, to render system (2.27) feedback passive can be summarized as determining a passivation law u_k and an output $\bar{y}_k$, such that relation (2.29) is satisfied with respect to the new input v_k.

2.5 DISCRETE-TIME HIGH ORDER NEURAL NETWORKS

The use of multilayer neural networks is well known for pattern recognition and for modeling of nonlinear functions. The NN is trained to learn an input-output map. Theoretical works have proven that, even with just one hidden layer, an NN can uniformly approximate any continuous function over a compact domain, provided that the NN has a sufficient number of synaptic connections.

For control tasks, extensions of the first order Hopfield model called Recurrent High Order Neural Networks (RHONN), which present more interactions among the neurons, are proposed in ([20], [23]). Additionally, the RHONN model is very flexible and allows us to incorporate in the neural model *a priori* information about the system structure.

Consider the following discrete-time recurrent high order neural network

(RHONN):

$$\widehat{x}_{i,k+1} = w_i^\top z_i(\widehat{x}_k, \upsilon_k), \quad i = 1, \cdots, n, \tag{2.33}$$

where $\widehat{x}_i$ $(i = 1, 2, \cdots, n)$ is the state of the i-th neuron, L_i is the respective number of high-order connections, $\{I_1, I_2, \cdots, I_{L_i}\}$ is a collection of non-ordered subsets of $\{1, 2, \cdots, n + m\}$, n is the state dimension, m is the number of external inputs, w_i $(i = 1, 2, \cdots, n)$ is the respective on-line adapted weight vector, and $z_i(\widehat{x}_k, \rho_k)$ is given by

$$z_i(x_k, \rho_k) = \begin{bmatrix} z_{i_1} \\ z_{i_2} \\ \vdots \\ z_{i_{L_i}} \end{bmatrix} = \begin{bmatrix} \prod_{j \in I_1} \xi_{i_j}^{d_{i_j}(1)} \\ \prod_{j \in I_2} \xi_{i_j}^{d_{i_j}(2)} \\ \vdots \\ \prod_{j \in I_{L_i}} \xi_{i_j}^{d_{i_j}(L_i)} \end{bmatrix}, \tag{2.34}$$

with $d_{j_i,k}$ being non-negative integers, and ξ_i defined as follows:

$$\xi_i = \begin{bmatrix} \xi_{i_1} \\ \vdots \\ \xi_{i_1} \\ \xi_{i_{n+1}} \\ \vdots \\ \xi_{i_{n+m}} \end{bmatrix} = \begin{bmatrix} S(x_1) \\ \vdots \\ S(x_n) \\ \rho_1 \\ \vdots \\ \rho_m \end{bmatrix}. \tag{2.35}$$

In (2.35), $\rho = [\rho_1, \rho_2, \ldots, \rho_m]^\top$ is the input vector to the neural network, and $S(\bullet)$ is defined by

$$S(\varsigma) = \frac{1}{1 + \exp(-\beta \varsigma)}, \quad \beta > 0, \tag{2.36}$$

where ς is any real value variable.

Consider the problem to approximating the general discrete-time nonlinear system (2.1), by the following discrete-time RHONN series-parallel representation [23]:

$$\widehat{x}_{i,k+1} = w_i^{*\top} z_i(x_k, \rho_k) + \varepsilon_{z_i}, \quad i = 1, \cdots, n, \tag{2.37}$$

where x_i is the i-th plant state, ε_{z_i} is a bounded approximation error, which can be reduced by increasing the number of the adjustable weights [23]. Assume that there exists an ideal weight vector w_i^* such that $\|\varepsilon_{z_i}\|$ can be minimized on a compact set $\Omega_{z_i} \subset \Re^{L_i}$. The ideal weight vector w_i^* is an artificial quantity required for analytical purposes [23]. In general, it is assumed that this vector exists and is constant but unknown. Let us define its estimate as w_i and the estimation error as

$$\widetilde{w}_{i,k} = w_i^* - w_{i,k}. \tag{2.38}$$

The estimate w_i is used for stability analysis, which will be discussed later. Since w_i^* is constant, then $\widetilde{w}_{i,k+1} - \widetilde{w}_{i,k} = w_{i,k+1} - w_{i,k}, \forall k \in 0 \cup \mathbb{Z}^+$.

From (2.33) three possible models can be derived:

- Parallel model

$$\widehat{x}_{i,k+1} = w_i^\top z_i(\widehat{x}_k, \rho_k), \quad i = 1, \cdots, n, \tag{2.39}$$

- Series-Parallel model

$$\widehat{x}_{i,k+1} = w_i^\top z_i(x_k, \rho_k), \quad i = 1, \cdots, n, \tag{2.40}$$

- Feedforward model (HONN)

$$\widehat{x}_{i,k} = w_i^\top z_i(\rho_k), \quad i = 1, \cdots, n, \tag{2.41}$$

where $\widehat{x}$ is the NN state vector, x is the plant state vector and ρ is the input vector to the NN.

2.6 THE EKF TRAINING ALGORITHM

The best well-known training approach for recurrent neural networks (RNNs) is back-propagation through time learning [28]. However, it is a first order gradient descent method and hence its learning speed could be very slow [16]. Recently, Extended Kalman Filter (EKF)-based algorithms have been introduced to train neural networks

[1, 7]. With the EKF-based algorithm, the learning convergence is improved [16]. The EKF training of neural networks, both feedforward and recurrent ones, has proven to be reliable and practical for many applications over the past ten years [7].

It is known that Kalman filtering (KF) estimates the state of a linear system with an additive state and an output of white noises [9, 26]. For EKF-based neural network training, the network weights become the states to be estimated. In this case, the error between the neural network output and the measured plant output can be considered as additive white noise. Due to the fact that the neural network mapping is nonlinear, an EKF-type is required (see [22] and references therein).

The training goal is to determine the optimal weight values which minimize the prediction error. The EKF-based training algorithm is described by [9]:

$$
\begin{aligned}
K_{i,k} &= P_{i,k}H_{i,k}\left[R_{i,k} + H_{i,k}^{\top}P_{i,k}H_{i,k}\right]^{-1} \\
w_{i,k+1} &= w_{i,k} + \eta_i K_{i,k}\left[y_k - \widehat{y}_k\right] \\
P_{i,k+1} &= P_{i,k} - K_{i,k}H_{i,k}^{\top}P_{i,k} + Q_{i,k}
\end{aligned}
\tag{2.42}
$$

where $P_i \in \mathfrak{R}^{L_i \times L_i}$ is the prediction error associated with the covariance matrix, $w_i \in \mathfrak{R}^{L_i}$ is the weight (state) vector, L_i is the total number of neural network weights, $y \in \mathfrak{R}^m$ is the measured output vector, $\widehat{y} \in \mathfrak{R}^m$ is the network output, η_i is a design parameter, $K_i \in \mathfrak{R}^{L_i \times m}$ is the Kalman gain matrix, $Q_i \in \mathfrak{R}^{L_i \times L_i}$ is the state noise associated covariance matrix, $R_i \in \mathfrak{R}^{m \times m}$ is the measurement noise associated covariance matrix, and $H_i \in \mathfrak{R}^{L_i \times m}$ is a matrix for which each entry (H_{ij}) is the derivative of one of the neural network outputs, $(\widehat{y})$, with respect to one neural network weight, (w_{ij}), as follows:

$$
H_{ij,k} = \left[\frac{\partial \widehat{y}_k}{\partial w_{ij,k}}\right]_{w_{i,k}=\widehat{w}_{i,k+1}}, \quad i = 1,\ldots,n \text{ and } j = 1,\ldots,L_i
\tag{2.43}
$$

Usually P_i, Q_i and R_i are initialized as diagonal matrices, with entries $P_i(0)$, $Q_i(0)$ and $R_i(0)$, respectively. It is important to note that $H_{i,k}$, $K_{i,k}$ and $P_{i,k}$ for the EKF are bounded [26]. Therefore, there exist constants $\overline{H}_i > 0$, $\overline{K}_i > 0$ and $\overline{P}_i > 0$ such that:

$$
\begin{aligned}
\|H_{i,k}\| &\leq \overline{H}_i \\
\|K_{i,k}\| &\leq \overline{K}_i \\
\|P_{i,k}\| &\leq \overline{P}_i
\end{aligned}
\tag{2.44}
$$

Comment 2.2 The measurement and process noises are typically characterized as zero-mean white noises with covariances given by $\delta_{k,j}R_{i,k}$ and $\delta_{k,j}Q_{i,k}$, respectively, with $\delta_{k,j}$ a Kronecker delta function (zero for $k \neq l$ and 1 for $k = l$) [10]. In order to simplify the notation in this book, the covariances will be represented by their respective associated matrices, $R_{i,k}$ and $Q_{i,k}$, for the noises and $P_{i,k}$ for the prediction error.

2.7 SEPARATION PRINCIPLE FOR DISCRETE-TIME NONLINEAR SYSTEMS

Consider a MIMO nonlinear system

$$
x_{k+1} = F(x_k, u_k) \tag{2.45}
$$

$$
y_k = h(x_k), \tag{2.46}
$$

where $x_k \in \mathbb{R}^n$, $k \in \mathbb{Z}^+ \cup 0 = \{0, 1, 2, \ldots\}$, $u_k \in \mathbb{R}^m$ is the input, and $F \in \mathbb{R}^n \times \mathbb{R}^m \to \mathbb{R}^n$ is a nonlinear function.

Theorem 2.4: (Separation Principle) [18]

The asymptotic stabilization problem of system (2.45), via estimated state feedback

$$u_k = \xi(\widehat{x}_k)$$
$$\widehat{x}_{k+1} = f(\widehat{x}_k) + g(\widehat{x}_k)u_k \tag{2.47}$$

is solvable, if and only if, the system (2.45) is asymptotically stabilizable and exponentially detectable. ∎

Corollary 2.1: [18]

There is an exponential observer for a Lyapunov stable discrete-time nonlinear system (2.45) with $u = 0$ if, and only if, the linear approximation

$$x_{k+1} = A_k x_k + B u_k$$
$$y_k = C x_k \tag{2.48}$$

$$A = \left.\frac{\partial f}{\partial x}\right|_{x=0}, \quad B = \left.\frac{\partial g}{\partial x}\right|_{x=0}, \quad C = \left.\frac{\partial h}{\partial x}\right|_{x=0}$$

of the system (2.45) is detectable.

REFERENCES

1. A. Y. Alanis. *Neural network training using Kalman Filtering*. Master's Dissertation, Cinvestav, Unidad Guadalajara, Guadalajara Jalisco Mexico, 2004 (in spanish).

2. G. L. Amicucci, S. Monaco, and D. Normand-Cyrot. Control Lyapunov stabilization of affine discrete-time systems. In *Proceedings of the 36th IEEE Conference on Decision and Control*, 1:923–924, San Diego, CA, USA, Dec 1997.

3. Z. Artstein. Stabilization with relaxed controls. *Nonlinear Analysis: Theory, Methods and Applications*, 7(11):1163–1173, 1983.

4. B. Brogliato, R. Lozano, B. Maschke, and O. Egeland. *Dissipative Systems Analysis and Control: Theory and Applications*. Springer-Verlag, Berlin, Germany, 2nd edition, 2007.

5. C. I. Byrnes and W. Lin. Losslessness, feedback equivalence, and the global stabilization of discrete-time nonlinear systems. *IEEE Transactions on Automatic Control*, 39(1):83–98, 1994.

6. C. Cruz-Hernandez, J. Alvarez-Gallegos, and R. Castro-Linares. Stability of discrete nonlinear systems under nonvanishing perturbations: application to a nonlinear model–matching problem. *IMA Journal of Mathematical Control & Information*, 16:23–41, 1999.

7. L. A. Feldkamp, D. V. Prokhorov and T. M. Feldkamp. Simple and conditioned adaptive behavior from Kalman filter trained recurrent networks. *Neural Networks*, 16:683–689, 2003.

8. R. A. Freeman and P. V. Kokotović. *Robust Nonlinear Control Design: State-Space and Lyapunov Techniques*. Birkhauser Boston Inc., Cambridge, MA, USA, 1996.

9. R. Grover and P. Y. C. Hwang. *Introduction to Random Signals and Applied Kalman Filtering, 2nd ed.* John Wiley and Sons, N. Y., USA, 1992.

10. S. Haykin. *Kalman Filtering and Neural Networks*. John Wiley and Sons, N. Y., USA, 2001.

11. C. M. Kellett and A. R. Teel. Results on discrete-time control-Lyapunov functions. In *Proceedings of the 42nd IEEE Conference on Decision and Control, 2003*, 6:5961–5966, Maui, Hawaii, USA, Dec 2003.

12. H. K. Khalil. *Nonlinear Systems*. Prentice-Hall, Upper Saddle River, NJ, USA, 1996.

13. D. E. Kirk. *Optimal Control Theory: An Introduction*. Prentice-Hall, Englewood Cliffs, NJ, USA, 1970.

14. M. Krstić and Z. Li. Inverse optimal design of input-to-state stabilizing nonlinear

controllers. *Automatic Control, IEEE Transactions on*, 43(3):336–350, 1998.

15. J. P. LaSalle. *The Stability and Control of Discrete Processes*. Springer-Verlag, Berlin, Germany, 1986.

16. C. Leunga, and L. Chan. Dual extended Kalman filtering in recurrent neural networks, *Neural Networks*, 16:223–239, 2003.

17. F. L. Lewis and V. L. Syrmos. *Optimal control*. Wiley, New York, U.S.A., 1995.

18. W. Lin. and C. I. Byrnes. Design of discrete-time nonlinear control systems via smooth feedback. *Automatic Control, IEEE Transactions on*, 39(11):2340–2346, 1994.

19. L. Magni and R. Scattolini. *Assessment and Future Directions of Nonlinear Model Predictive Control*, volume 358 of *Lecture Notes in Control and Information Sciences*. Springer-Verlag, Berlin, Germany, 2007.

20. K. S. Narendra and K. Parthasarathy. Identification and control of dynamical systems using neural networks. *IEEE Transactions on Neural Networks*, 1:4–27, Mar. 1990.

21. Z. Ping-Jiang, E. D. Sontag, and Y. Wang. Input-to-state stability for discrete-time nonlinear systems. *Automatica*, 37:857–869, 1999.

22. A. S. Poznyak, E. N. Sanchez and W. Yu. *Differential Neural Networks for Robust Nonlinear Control* World Scientific, Singapore, 2001.

23. G. A. Rovithakis and M. A. Chistodoulou. *Adaptive Control with Recurrent High-Order Neural Networks* Springer Verlag, Berlin, Germany, 2000.

24. P. O. M. Scokaert, J. B. Rawlings, and E. S. Meadows. Discrete-time stability with perturbations: Application to model predictive control. *Automatica*, 33(3):463–470, 1997.

25. R. Sepulchre, M. Jankovic, and P. V. Kokotović. *Constructive Nonlinear Control*. Springer-Verlag, Berlin, Germany, 1997.

26. Y. Song and J. W. Grizzle. The extended Kalman filter as local asymptotic observer for discrete-time nonlinear systems *Journal of Mathematical Systems, Estimation and Control*, 5(1): 59-78,1995.

27. M. Vidyasagar. *Nonlinear Systems Analysis*. Prentice-Hall, Englewood Cliffs,

NJ, USA, 2nd edition, 1993.

28. R. J. Williams and D. Zipser, A learning algorithm for continually running fully recurrent neural networks. *Neural Computation*, 1:270–280, 1989.

3 Discrete-Time Neural Block Control

This chapter deals with adaptive trajectory tracking for a class of MIMO discrete-time nonlinear systems in the presence of bounded disturbances. A recurrent high order neural network is first used to identify the plant model, then based on this neural model, a discrete-time control law, which combines discrete-time block control and sliding mode techniques, is derived. The chapter also includes the respective stability analysis for the whole system. A strategy is also proposed to avoid zero-crossing of specific adaptive weights. Applicability of the proposed scheme is illustrated via simulation of a discrete-time nonlinear controller for an induction motor.

Frequently, modern control schemes for nonlinear systems require a very structured knowledge about the system to be controlled; such knowledge should be represented in terms of differential or difference equations. This mathematical description of the dynamic system is called the model. Basically there are two ways to obtain a model; it can be derived in a deductive manner using physics laws, or it can be inferred from a set of data collected during a practical experiment. The first method can be simple, but in many cases it is excessively time-consuming; sometimes, it would be unrealistic or impossible to obtain an accurate model in this way. The second method, which is commonly referred to as system identification, could be a useful shortcut for deriving mathematical models. Although system identification does not always result in an equally accurate model, a satisfactory one can be often obtained with reasonable efforts. The main drawback is the requirement to conduct a practical experiment which brings the system through its range of operation. Besides, a certain knowledge about the plant is still required. Once such a model obtained, based on it, the next step is to synthesize an adequate control law to obtain the specified objectives.

3.1 IDENTIFICATION

In this section, Let us consider the problem of identifying the nonlinear system

$$x_{k+1} = F(x_k, u_k), \tag{3.1}$$

where $x \in \Re^n$, $u \in \Re^m$ and $F \in \Re^n \times \Re^m \to \Re^n$ is a nonlinear function. Now, to identify system (3.1) we use a series-parallel RHONN defined as:

$$\widehat{x}_{i,k+1} = w_i^\top z_i(x_k, u_k), \quad i = 1, \cdots, n, \tag{3.2}$$

where $\widehat{x}_i$ $(i = 1, 2, \cdots, n)$ is the state of the i-th neuron, L_i is the respective number of high-order connections, $\{I_1, I_2, \cdots, I_{L_i}\}$ is a collection of non-ordered subsets of $\{1, 2, \cdots, n+m\}$, n is the state dimension, m is the number of external inputs, w_i $(i = 1, 2, \cdots, n)$ is the respective on-line adapted weight vector, with $z_i(x_k, u_k)$ as defined in (3.20).

Consider the problem of approximating the general discrete-time nonlinear system (3.1), by the following discrete-time RHONN series-parallel representation [3]:

$$\widehat{x}_{i,k+1} = w_i^{*\top} z_i(x_k, u_k) + \varepsilon_{z_i}, \quad i = 1, \cdots, n, \tag{3.3}$$

where x_i is the i-th plant state, and ε_{z_i} is a bounded approximation error, which can be reduced by increasing the number of the adjustable weights [3]. Assume that there exists an ideal weight vector w_i^* such that $\|\varepsilon_{z_i}\|$ can be minimized on a compact set $\Omega_{z_i} \subset \Re^{L_i}$. The ideal weight vector w_i^* is an artificial quantity required for analytical purposes [3]. In general, it is assumed that this vector exists and is constant but unknown. Let us define its estimate as w_i and the estimation error as

$$\widetilde{w}_{i,k} = w_i^* - w_{i,k}. \tag{3.4}$$

The estimate w_i is used for the stability analysis, which will be discussed later. Since w_i^* is constant, then

$$\widetilde{w}_{i,k+1} - \widetilde{w}_{i,k} = w_{i,k+1} - w_{i,k}, \forall k \in 0 \cup \mathbb{Z}^+.$$

The RHONN is trained with a modified Extended Kalman Filter (EKF) algorithm defined by:

$$
\begin{aligned}
w_{i,k+1} &= w_{i,k} + \eta_i K_{i,k} e_{i,k} & (3.5)\\
K_{i,k} &= \begin{cases} P_{i,k} H_{i,k} M_{i,k} & \text{if } \|w_{i,k}\| > c_i \\ 0 & \text{if } \|w_{i,k}\| < c_i \end{cases} \\
P_{i,k+1} &= P_{i,k} - K_{i,k} H_{i,k}^\top P_{i,k} + Q_{i,k} \\
i &= 1, \cdots, n
\end{aligned}
$$

with

$$M_{i,k} = \left[R_{i,k} + H_{i,k}^\top P_{i,k} H_{i,k} \right]^{-1} \tag{3.6}$$

$$e_{i,k} = x_{i,k} - \widehat{x}_{i,k}, \tag{3.7}$$

where $c_i > 0$ is a constraint used to avoid the zero-crossing, $e_{i,k} \in \mathfrak{R}$ is the respective identification error, $P_{i,k} \in \mathfrak{R}^{L_i \times L_i}$ is the prediction error associated covariance matrix at step k, $w_i \in \mathfrak{R}^{L_i}$ is the weight (state) vector, L_i is the respective number of neural network weights, x_i is the i-th plant state, $\widehat{x}_i$ is the i-th neural network state, n is the number of states, $K_i \in \mathfrak{R}^{L_i}$ is the Kalman gain vector, $Q_i \in \mathfrak{R}^{L_i \times L_i}$ is the state noise associated covariance matrix, $R_i \in \mathfrak{R}$ is the measurement noise associated covariance; $H_i \in \mathfrak{R}^{L_i}$ is a vector in which each entry (H_{ij}) is the derivative of one of the neural network states, $(\widehat{x}_i)$, with respect to one neural network weight, (w_{ij}), defined as follows

$$H_{ij,k} = \left[\frac{\partial \widehat{x}_{i,k}}{\partial w_{ij,k}} \right]^\top_{w_{i,k} = w_{i,k+1}}, \tag{3.8}$$

where $i = 1, ..., n$ and $j = 1, ..., L_i$. If we select $c_i = 0$ the modified EKF (3.5) becomes the standard extended Kalman Filter [6]. Usually P_i and Q_i are initialized as diagonal matrices, with entries $P_i(0)$ and $Q_i(0)$, respectively. It is important to remark that $H_{i,k}$, $K_{i,k}$ and $P_{i,k}$ for the EKF are bounded; for a detailed explanation of this fact see [4].

Then the dynamics of (3.7) can be expressed as

$$e_{i,k+1} = \widetilde{w}_{i,k} z_i(x_k, u_k) + \varepsilon_{z_i}. \tag{3.9}$$

On the other hand, the dynamics of (3.4) is

$$\widetilde{w}_{i,k+1} = \widetilde{w}_{i,k} - \eta_i K_{i,k} e_k. \tag{3.10}$$

Now, we establish the first main result of this chapter in the following theorem.

Theorem 3.1

The RHONN (3.2) trained with the modified EKF-based algorithm (3.5) to identify the nonlinear plant (3.1), ensures that the identification error (3.7) is semiglobally uniformly ultimately bounded (SGUUB); moreover, the RHONN weights remain bounded. ∎

Proof

Case 1. $\|w_{i,k}\| > c_i$. Consider the Lyapunov function candidate, for $i = 1, 2, \ldots, n$

$$
\begin{aligned}
V_{i,k} &= \widetilde{w}_{i,k}^T P_{i,k} \widetilde{w}_{i,k} + e_{i,k}^2 \\
\Delta V_{i,k} &= V_{k+1} - V_k \\
&= \widetilde{w}_{i,k+1}^T P_{i,k+1} \widetilde{w}_{i,k+1} + e_{i,k+1}^2 \\
&\quad - \widetilde{w}_{i,k}^T P_{i,k} \widetilde{w}_{i,k} - e_{i,k}^2.
\end{aligned}
\tag{3.11}
$$

Using (3.9) and (3.10) in (3.11)

$$
\begin{aligned}
\Delta V_{i,k} &= \left[\widetilde{w}_{i,k} - \eta_i K_{i,k} e_{i,k} \right]^T \\
&\quad \times [P_{i,k} - A_{i,k}] \left[\widetilde{w}_{i,k} - \eta_i K_{i,k} e_{i,k} \right] \\
&\quad + \left[\widetilde{w}_{i,k} z_i(x_k, u_k) + \varepsilon_{z_i} \right]^2 \\
&\quad - \widetilde{w}_{i,k} P_{i,k} \widetilde{w}_{i,k} - e_{i,k}^2,
\end{aligned}
\tag{3.12}
$$

with $A_{i,k} = K_{i,k} H_{i,k}^\top P_{i,k} + Q_{i,k}$; then, (3.12) can be expressed as

$$
\begin{aligned}
\Delta V_{i,k} &= \widetilde{w}_{i,k}^T P_{i,k} \widetilde{w}_{i,k} - \eta e_{i,k} K_{i,k}^T P_{i,k} \widetilde{w}_{i,k} - \widetilde{w}_{i,k}^T A_{i,k} \widetilde{w}_{i,k} + \eta e_{i,k} K_{i,k}^T A_{i,k} \widetilde{w}_{i,k} \\
&\quad - \eta e_{i,k} \widetilde{w}_{i,k}^T P_{i,k} K_{i,k} + \eta^2 e_{i,k}^2 K_{i,k}^T P_{i,k} K_{i,k} + \eta e_{i,k} \widetilde{w}_{i,k}^T A_{i,k} K_{i,k} - \eta^2 e_{i,k}^2 K_{i,k}^T A_{i,k} K_{i,k} \\
&\quad + \left(\widetilde{w}_{i,k} z_i(x_k, u_k) \right)^2 + 2\varepsilon_{z_i} \widetilde{w}_{i,k} z_i(x_k, u_k) + \varepsilon_{z_i}^2 - \widetilde{w}_{i,k}^T P_{i,k} \widetilde{w}_{i,k} - e_{i,k}^2.
\end{aligned}
\tag{3.13}
$$

Using the inequalities

$$
\begin{aligned}
X^T X + Y^T Y &\geq 2 X^T Y \\
X^T X + Y^T Y &\geq -2 X^T Y \\
-\lambda_{\min}(P) X^2 &\geq -X^T P X \geq -\lambda_{\max}(P) X^2,
\end{aligned}
\tag{3.14}
$$

which are valid $\forall X, Y \in \Re^n$, $\forall P \in \Re^{n \times n}$, $P = P^T > 0$, then (3.13) can be rewritten as

$$
\begin{aligned}
\Delta V_{i,k} \leq & -\widetilde{w}_{i,k}^T A_{i,k} \widetilde{w}_{i,k} - \eta^2 e_{i,k}^2 K_{i,k}^T A_{i,k} K_{i,k} + \widetilde{w}_{i,k}^T \widetilde{w}_{i,k} + e_{i,k}^2 \\
& + \eta^2 e_{i,k}^2 K_{i,k}^T P_{i,k} P_{i,k}^T K_{i,k} + \eta^2 \widetilde{w}_i^T A_{i,k} K_{i,k} K_{i,k}^T A_{i,k}^T \widetilde{w}_{i,k} \\
& + \eta^2 e_{i,k}^2 K_{i,k}^T P_{i,k} K_{i,k} + 2 \left(\widetilde{w}_{i,k} z_i \left(x_k, u_k \right) \right)^2 + 2 \varepsilon_{z_i}^2 - e_{i,k}^2. \quad (3.15)
\end{aligned}
$$

Then

$$
\begin{aligned}
\Delta V_{i,k} \leq & -\left\| \widetilde{w}_{i,k} \right\|^2 \lambda_{\min} \left(A_{i,k} \right) - \eta^2 e_{i,k}^2 \left\| K_{i,k} \right\|^2 \lambda_{\min} \left(A_{i,k} \right) + \left\| \widetilde{w}_{i,k} \right\|^2 \\
& + \eta^2 e_{i,k}^2 \left\| K_{i,k} \right\|^2 \lambda_{\max}^2 \left(P_{i,k} \right) + \eta^2 \left\| \widetilde{w}_{i,k} \right\|^2 \lambda_{\max}^2 \left(A_{i,k} \right) \left\| K_{i,k} \right\|^2 \\
& + \eta^2 e_{i,k}^2 \left\| K_{i,k} \right\|^2 \lambda_{\max} \left(P_{i,k} \right) + 2 \left\| \widetilde{w}_{i,k} \right\|^2 \left\| z_i \left(x_k, u_k \right) \right\|^2 + 2 \varepsilon_{z_i}^2. \quad (3.16)
\end{aligned}
$$

Defining

$$
\begin{aligned}
E_{i,k} &= \lambda_{\min} \left(A_{i,k} \right) - \eta^2 \lambda_{\max}^2 \left(A_{i,k} \right) \left\| K_{i,k} \right\|^2 - 2 \left\| z_i \left(x_k, u_k \right) \right\|^2 - 1 \\
F_{i,k} &= \eta^2 \left\| K_{i,k} \right\|^2 \lambda_{\min} \left(A_{i,k} \right) - \eta^2 \left\| K_{i,k} \right\|^2 \lambda_{\max}^2 \left(P_{i,k} \right) - \eta^2 \left\| K_{i,k} \right\|^2 \lambda_{\max} \left(P_{i,k} \right),
\end{aligned}
$$

and selecting η_i, Q_i and R_i, such that $E_i > 0$ and $F_i > 0$, $\forall k$, then (3.16) can be expressed as

$$
\Delta V_{i,k} \leq - \left\| \widetilde{w}_{i,k} \right\|^2 E_{i,k} - \left| e_{i,k} \right|^2 F_{i,k} + 2 \varepsilon_{z_i}^2.
$$

Hence $\Delta V_{i,k} < 0$ when

$$
\left\| \widetilde{w}_{i,k} \right\| > \frac{\sqrt{2} \left| \varepsilon_{z_i} \right|}{\sqrt{E_{i,k}}} \equiv \kappa_1 \quad (3.17)
$$

and

$$
\left| e_{i,k} \right| > \frac{\sqrt{2} \left| \varepsilon_{z_i} \right|}{\sqrt{F_{i,k}}} \equiv \kappa_2. \quad (3.18)
$$

Therefore, the solution of (3.9) and (3.10) are SGUUB.

Case 2. $\left\| w_{i,k} \right\| < c_i$. Consider the same Lyapunov function candidate as in Case 1 (3.10). Following the same procedure with $K_i = 0$, then $\Delta V_{i,k} < 0$ when (3.18) is fulfilled; hence, as in Case 1, the solution of (3.9) and (3.10) are SGUUB. ∎

Comment 3.1 As well as many feedback linearization-like controllers [1], the neural block controller may present some singularities, due to the zero crossing of some adaptive parameters. To overcome the controller singularity problem, this chapter includes the constraint c_i, which allows us to eliminate singularities for specific weights zero-crossing [1].

3.2 ILLUSTRATIVE EXAMPLE

In this section we apply the above developed scheme to a three-phase induction motor model. The six order discrete-time induction motor model in the stator fixed reference frame (α, β) under the assumptions of equal mutual inductances and linear magnetic circuit is given by [2]

$$
\begin{aligned}
\omega_{k+1} &= \omega_k + \frac{\mu}{\alpha}(1-\alpha)M\left(i_k^\beta \psi_k^\alpha - i_k^\alpha \psi_k^\beta\right) - \left(\frac{T}{J}\right)T_{L,k}\\
\psi_{k+1}^\alpha &= \cos(n_p\theta_{k+1})\rho_{1,k} - \sin(n_p\theta_{k+1})\rho_{2,k}\\
\psi_{k+1}^\beta &= \sin(n_p\theta_{k+1})\rho_{1,k} + \cos(n_p\theta_{k+1})\rho_{2,k}\\
i_{k+1}^\alpha &= \varphi_k^\alpha + \frac{T}{\sigma}u_k^\alpha\\
i_{k+1}^\beta &= \varphi_k^\beta + \frac{T}{\sigma}u_k^\beta\\
\theta_{k+1} &= \theta_k + \omega_k T + \frac{\mu}{\alpha}\left[T - \frac{(1-a)}{\alpha}\right]\\
&\quad \times M\left(i_k^\beta \psi_k^\alpha - i_k^\alpha \psi_k^\beta\right) - \frac{T_{L,k}}{J}T^2,
\end{aligned}
\tag{3.19}
$$

with

$$\rho_{1,k} = a\left(\cos(\phi_k)\psi_k^\alpha + \sin(n_p\phi_k)\psi_k^\beta\right)$$
$$+b\left(\cos(\phi_k)i_k^\alpha + \sin(\phi_k)i_k^\beta\right)$$
$$\rho_{2,k} = a\left(\cos(\phi_k)\psi_k^\alpha - \sin(\phi_k)\psi_k^\beta\right)$$
$$+b\left(\cos(\phi_k)i_k^\alpha - \sin(\phi_k)i_k^\beta\right)$$
$$\varphi_k^\alpha = i_k^\alpha + \alpha\beta T\psi_k^\alpha + n_p\beta T\omega_k\psi_k^\alpha - \gamma T i_k^\alpha$$
$$\varphi_k^\beta = i_k^\beta + \alpha\beta T\psi_k^\beta + n_p\beta T\omega_k\psi_k^\beta - \gamma T i_k^\beta$$
$$\phi_k = n_p\theta_k, \tag{3.20}$$

with $b = (1-a)M$, $\alpha = \frac{R_r}{L_r}$, $\gamma = \frac{M^2 R_r}{\sigma L_r^2} + \frac{R_s}{\sigma}$, $\sigma = L_s - \frac{M^2}{L_r}$, $\beta = \frac{M}{\sigma L_r}$, $a = e^{-\alpha T}$ and $\mu = \frac{Mn_p}{JL_r}$; besides Ls, L_r and M are the stator, rotor and mutual inductance respectively; R_s and R_r are the stator and rotor resistances respectively; n_p is the number of pole pairs; i^α and i^β represent the currents in the α and β phases, respectively; ψ^α and ψ^β represent the fluxes in the α and β phases, respectively; and θ is the rotor angular displacement. Simulations are performed for the system (3.19), using the following parameters: $R_s = 14\Omega$; $L_s = 400mH$; $M = 377mH$; $R_r = 10.1\Omega$; $L_r = 412.8mH$; $n_p = 2$; $J = 0.01Kgm^2$; $T = 0.001s$.

The RHONN proposed for this application is as follows:

$$x_{1,k+1} = w_{11,k}S(\omega_k) + w_{12,k}S(\omega)S\left(\psi_k^\beta\right)i_k^\alpha$$
$$+w_{13,k}S(\omega)S(\psi_k^\alpha)i_k^\beta$$
$$x_{2,k+1} = w_{21,k}S(\omega_k)S\left(\psi_k^\beta\right) + w_{22,k}i_k^\beta$$
$$x_{3,k+1} = w_{31,k}S(\omega_k)S(\psi_k^\alpha) + w_{32,k}i_k^\alpha$$
$$x_{4,k+1} = w_{41,k}S(\psi_k^\alpha) + w_{42,k}S\left(\psi_k^\beta\right)$$
$$+w_{43,k}S(i_k^\alpha) + w_{44,k}u_k^\alpha$$
$$x_{5,k+1} = w_{51,k}S(\psi_k^\alpha) + w_{52,k}S\left(\psi_k^\beta\right)$$
$$+w_{53,k}S\left(i_k^\beta\right) + w_{54,k}u_k^\beta.$$

The training is performed on-line, using a series-parallel configuration. During the identification process, the plant and the NN operate in an open loop. Both of them (plant and NN) have the same input vector $\begin{bmatrix} u_\alpha, & u_\beta \end{bmatrix}^\top$; u_α and u_β are chirp functions with 170 volts of maximal amplitude and incremental frequencies from 0 Hz to 250 Hz and 0 Hz to 200 Hz respectively. All the NN states are initialized in a random way as well as the weight vectors. It is important to remark that the initial conditions of the plant are completely different from the initial conditions for the NN. The identification is performed using (2.41) with $i = 1, 2, \cdots, n$ with n the dimension of plant states $(n = 6)$.

The results of the simulation are presented in Figs. 3.1–3.5. Fig. 3.1 displays the identification performance for the speed rotor; Fig. 3.2 and Fig. 3.3 present the identification performance for the fluxes in phase α and β, respectively. Figs 3.4 and 3.5 portray the identification performance for currents in phase α and β, respectively.

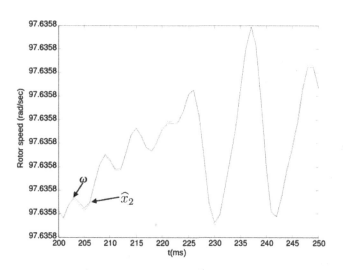

FIGURE 3.1   Rotor speed identification.

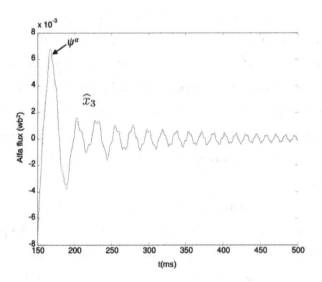

FIGURE 3.2 ψ^α identification.

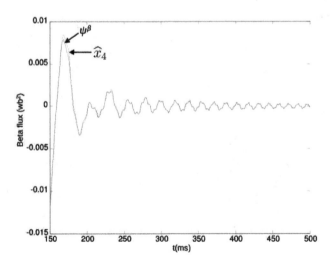

FIGURE 3.3 ψ^β identification.

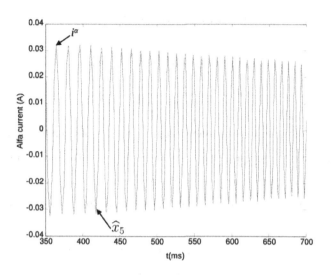

FIGURE 3.4 i^α identification.

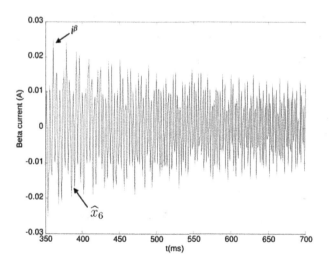

FIGURE 3.5 i^β identification.

3.3 NEURAL BLOCK CONTROLLER DESIGN

Consider the following special case of system (3.1):

$$
\begin{aligned}
x_{k+1} &= f(x_k) + B(x_k)u_k + d_k \\
y_k &= Cx_k,
\end{aligned}
\tag{3.21}
$$

where $x \in \Re^n$ is the state vector of the system, $u_k \in \Re^m$ is the input vector, $y_k \in \Re^p$ is the output vector, the vector $f(\cdot)$, the columns of $B(\cdot)$ and $d(\cdot)$ are smooth vector fields, and $d(\cdot)$ is a disturbance vector. By means of non-singular transformation ([2, 5]), system (3.21) can be represented in the block controllable form as follows:

$$
\begin{aligned}
x_{i,k+1} &= f_i(\bar{x}_{i,k}) + B_i(\bar{x}_{i,k})x_{i+1,k} + d_{i,k} \\
x_{r,k+1} &= f_r(x_k) + B_r(x_k)u_k + d_{r,k} \\
y_k &= x_{1,k}, \; i = 1, \cdots, r-1,
\end{aligned}
\tag{3.22}
$$

where $x_k = \begin{bmatrix} x_{1,k} & \cdots & x_{i,k} & \cdots & x_{r,k} \end{bmatrix}^\top$, $\bar{x}_{i,k} = \begin{bmatrix} x_{1,k} & \cdots & x_{i,k} \end{bmatrix}^\top$ and $d_k = \begin{bmatrix} d_{1,k} & \cdots & d_{i,k} & \cdots & d_{r,k} \end{bmatrix}^\top$, $i = 1, \cdots, r-1$, and the set of numbers $(n_1, \cdots, n_r)$, which define the structure of system (3.22), satisfy $n_1 \le n_2 \le \cdots \le n_r \le m$.

Define the following transformation:

$$
\begin{aligned}
\mathbf{z}_{1,k} &= x_{1,k} - x_{1,k}^d \\
\mathbf{z}_{2,k} &= x_{2,k} - x_{2,k}^d \\
&= x_{2,k} - \left[B_1(x_{1,k}) \right]^{-1} \left(\mathbf{K}_1 \mathbf{z}_{1,k} - \left(f_1(x_{1,k}) - d_{1,k} \right) \right) \\
\mathbf{z}_{3,k} &= x_{3,k} - x_{3,k}^d \\
&= x_{3,k} - \left[B_2(x_{2,k}) \right]^{-1} \left(\mathbf{K}_2 \mathbf{z}_{2,k} - \left(f_2(x_{2,k}) - d_{2,k} \right) \right) \\
&\quad\vdots \\
\mathbf{z}_{r,k} &= x_{r,k} - x_{r,k}^d,
\end{aligned}
\tag{3.23}
$$

where $y_{d,k} = x_{1,k}^d$ is the desired trajectory for tracking, x_i^d is the desired value for x_i

$(i = 1, 2, \cdots, r)$, which will be defined later, and $\mathbf{K}_i$ is a Schur matrix. Using (3.23), system (3.22) can be rewritten as

$$z_{1,k+1} = \mathbf{K}_1 z_{1,k} + B_1 z_{2,k}$$
$$\vdots$$
$$z_{r-1,k+1} = \mathbf{K}_{r-1} z_{r-1,k} + B_{r-1} z_{r,k}$$
$$z_{r,k+1} = f_r(x_k) + B_r(x_k) u_k + d_{r,k} - x^d_{r,k+1}. \tag{3.24}$$

To design the control law, we use the sliding mode block control technique. The manifold can be derived from the block control structure, and a natural selection is $S_{D,k} = \mathbf{z}_{r,k} = 0$. Thus, system (3.24) is represented, in the new variables, as

$$\mathbf{z}_{1,k+1} = \mathbf{K}_1 \mathbf{z}_{1,k} + B_1 \mathbf{z}_{2,k}$$
$$\vdots$$
$$\mathbf{z}_{r-1,k+1} = \mathbf{K}_{r-1} \mathbf{z}_{r-1,k} + B_{r-1} S_{D,k}$$
$$S_{D,k+1} = f_r(x_k) + B_r(x_k) u_k + d_{r,k} - x^d_{r,k+1}. \tag{3.25}$$

Once the sliding manifold is selected, the next step is to define $u(k)$, as

$$u_k = \begin{cases} u_{eq,k} & \text{for} \ \ \|u_{eq,k}\| \le u_0 \\ u_0 \dfrac{u_{eq,k}}{\|u_{eq,k}\|} & \text{for} \ \ \|u_{eq,k}\| > u_0 \end{cases}, \tag{3.26}$$

where the equivalent control is calculated from $S_{D,k+1} = 0$, as

$$u_{eq,k} = [B_r(x_k)]^{-1} \left(-f_r(x_k) + x^d_{r,k+1} - d_{r,k} \right).$$

To this end, we present a stability analysis to prove that the closed-loop system (3.25)–(3.26) motion over the manifold is stable, which is the second main result of this chapter.

Theorem 3.2

The control law (3.26) ensures the sliding manifold $S_{D,k} = z_{r,k} = 0$ is stable, for system (3.22). ∎

Proof

Write the last subsystem of (3.25) as

$$S_{D,k+1} = S_{D,k} - x_{r,k} + x_{r,k}^d + f_r\left(x_k^1\right) + B_r u_k + d_{r,k} - x_{r,k+1}^d.$$

Note that when $\|u_{eq,k}\| \le u_0$, the equivalent control is applied yielding motion on the sliding manifold $S_{D,k} = 0$. For the case of $\|u_{eq,k}\| > u_0$, the proposed control strategy is $u_0 \frac{u_{eq,k}}{\|u_{eq,k}\|}$, and the closed-loop system is

$$
\begin{aligned}
S_{D,k+1} &= S_{D,k} - x_{r,k} + x_{r,k}^d + f_r\left(x_k^1\right) + B_r u_0 \frac{u_{eq,k}}{\|u_{eq,k}\|} \\
&\quad + d_{r,k} - x_{r,k+1}^d \\
&= \left(S_{D,k} - x_{r,k} + x_{r,k}^d + f_r\left(x_k^1\right) + d_{r,k} - x_{r,k+1}^d\right) \\
&\quad \times \left(1 - \frac{u_0}{\|u_{eq,k}\|}\right).
\end{aligned}
$$

Along any solution of the system, the Lyapunov candidate function $V_k = S_{D,k}^\top S_{D,k}$

gives

$$\Delta V_k = S_{D,k+1}^\top S_{D,k+1} - S_{D,k}^\top S_{D,k}$$

$$= \left[(S_{D,k} + f_{s,k}) \left(1 - \frac{u_0}{\|u_{eq,k}\|} \right) \right]^\top$$

$$\times (S_{D,k} + f_{s,k}) \left(1 - \frac{u_0}{\|u_{eq,k}\|} \right) - S_{D,k}^\top S_{D,k}$$

$$\le \left[\|S_{D,k} + f_{s,k}\| \left(1 - \frac{u_0}{\|u_{eq,k}\|} \right) \right]^\top$$

$$\times \|S_{D,k} + f_{s,k}\| \left(1 - \frac{u_0}{\|u_{eq,k}\|} \right) - \|S_{D,k}\|^2 .$$

Then

$$\Delta V_k \le \left[\|S_{D,k} + f_{s,k}\| - \frac{u_0}{\|B_r^{-1}\|} \right]^\top \left(\|S_{D,k} + f_{s,k}\| - \frac{u_0}{\|B_r^{-1}\|} \right)$$

$$- \|S_{D,k}\|^2$$

$$\le \left(\|S_{D,k} + f_{s,k}\| - \frac{u_0}{\|B_r^{-1}\|} \right)^2 - \|S_{D,k}\|^2$$

$$\le \|S_{D,k}\|^2 + 2\|S_{D,k}\|^2 f_{s,k} - 2u_0 \frac{\|S_{D,k}\|}{\|B_r^{-1}\|} + \|f_{s,k}\|^2$$

$$- 2u_0 \frac{\|f_{s,k}\|}{\|B_r^{-1}\|} + \frac{u_0^2}{\|B_r^{-1}\|^2} - \|S_{D,k}\|^2$$

$$\le -2\|S_{D,k}\| \left(\frac{u_0}{\|B_r^{-1}\|} - \|f_{s,k}\| \right) + \left(\frac{u_0}{\|B_r^{-1}\|} - \|f_{s,k}\| \right)^2 ,$$

where $f_{s,k} = -x_{r,k} + x_{d,k} + f_r(x_k) + d_{r,k} - x_{r,k+1}^d$, and if $\|B_r^{-1}\| \|f_{s,k}\| \le u_0 \le \|B_r^{-1}\| (2\|S_{D,k}\| + \|f_{s,k}\|)$ holds, then $\Delta V_k \le 0$ [2], hence $\|S_{D,k}\|$ and $\|u_{eq,k}\|$ both decrease monotonically. Note that $\|B_r^{-1}\| (2\|S_{D,k}\| + \|f_{s,k}\|) \ge u_0$ is a greater bound than that established by the current case, i.e., $\|u_{eq,k}\| > u_0$, due to the fact that $\|B_r^{-1}\| (2\|S_{D,k}\| + \|f_{s,k}\|) \ge \|u_{eq,k}\|$. Therefore, the only condition drawn from the Lyapunov analysis is: $\|B_r^{-1}\| \|f_{s,k}\| \le u_0$ [2]. ∎

The proposed control scheme is shown in Figure 3.6. To this end, the third main result of this chapter is the following one.

Proposition 3.1

Given a desired output trajectory $y_d = x_r^d$, a dynamical system with output y, and a neural network with output $\hat{y}$, then it is possible to establish the following inequality [1]:

$$\|y_d - y\| \leq \|\hat{y} - y\| + \|y_d - \hat{y}\|, \tag{3.27}$$

where $y_d - y$ is the system output tracking error, $\hat{y} - y$ is the output identification error, and $y_d - \hat{y}$ is the RHONN output tracking error.

Based on this proposition, it is possible to divide the tracking error in two parts [1]:

1. Minimization of $\hat{y} - y$, which can be achieved by the proposed on-line iden-tification algorithm (3.1) on the basis of *Theorem* 3.1.
2. Minimization of $y_d - \hat{y}$, for which a tracking algorithm is developed on the basis of the neural identifier (3.2). This can be reached by designing a control law based on the RHONN model. To design such a controller, we propose to use the NBC methodology [1, 2].

Comment 3.2 It is possible to establish Proposition 3.1 by applying the separation principle for discrete-time nonlinear systems.

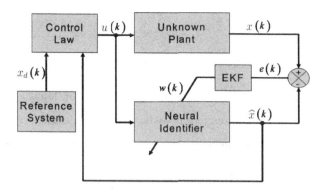

FIGURE 3.6   Neural block control scheme.

3.4 APPLICATIONS

In this section we apply the above developed scheme (Figure 3.6) to control a three-phase induction motor, whose model is described above.

3.4.1 NEURAL NETWORK IDENTIFICATION

The RHONN proposed for this application is as follows:

$$
\begin{aligned}
\widehat{x}_{1,k+1} &= w_{11,k}S(\omega_k) + w_{12,k}S(\omega)S\left(\psi_k^\beta\right)i_k^\alpha \\
&\quad + w_{13,k}S(\omega)S(\psi_k^\alpha)i_k^\beta \\
\widehat{x}_{2,k+1} &= w_{21,k}S(\omega_k)S\left(\psi_k^\beta\right) + w_{22,k}i_k^\beta \\
\widehat{x}_{3,k+1} &= w_{31,k}S(\omega_k)S(\psi_k^\alpha) + w_{32,k}i_k^\alpha \\
\widehat{x}_{4,k+1} &= w_{41,k}S(\psi_k^\alpha) + w_{42,k}S\left(\psi_k^\beta\right) \\
&\quad + w_{43,k}S(i_k^\alpha) + w_{44,k}u_k^\alpha \\
\widehat{x}_{5,k+1} &= w_{51,k}S(\psi_k^\alpha) + w_{52,k}S\left(\psi_k^\beta\right) \\
&\quad + w_{53,k}S\left(i_k^\beta\right) + w_{54,k}u_k^\beta.
\end{aligned}
\tag{3.28}
$$

The training is performed on-line, using a series-parallel configuration. All the NN states are initialized in a random way as well as the weights vectors. It is important to note that the initial conditions of the plant are completely different from the initial conditions for the NN.

3.4.2 NEURAL BLOCK CONTROLLER DESIGN

Given full state measurements, the control objective is to develop velocity and flux amplitude tracking for the discrete-time induction motor model (3.28), using the discrete-time control algorithm developed above. Let us define the following states as

$$
x_k^1 = \begin{bmatrix} x_{1,k} - \omega_{r,k} \\ \Psi_k - \Psi_{r,k} \end{bmatrix}, \quad x_k^2 = \begin{bmatrix} i_k^\alpha \\ i_k^\beta \end{bmatrix}, \tag{3.29}
$$

where $\Psi_k = \hat{x}_{2,k}^2 + \hat{x}_{3,k}^2$ is the rotor flux identified magnitude, $\Psi_{r,k}$ and $\omega_{r,k}$ are reference signals. Then

$$
\begin{aligned}
\Psi_{k+1} &= w_{21,k}^2 S^2(\omega_k) S^2\left(\psi_k^\beta\right) + w_{22,k}^2 i_k^{\beta^2} + w_{32,k}^2 i_k^{\alpha^2} \\
&+ w_{31,k}^2 S^2(\omega_k) S^2(\psi_k^\alpha) \\
&+ 2 w_{21,k} S(\omega_k) S\left(\psi_k^\beta\right) w_{22,k} i_k^\beta \\
&+ 2 w_{31,k} S(\omega_k) S(\psi_k^\alpha) w_{32,k} i_k^\alpha.
\end{aligned}
$$

Using (3.28), (3.29) can be represented in the block control form consisting of two blocks

$$
\begin{aligned}
x_{k+1}^1 &= f_1\left(x_k^1\right) + B_1\left(x_k^1\right) x_k^2 \\
x_{k+1}^2 &= f_2\left(x_k^1, x_k^2\right) + B_{2,k} u_k
\end{aligned} \tag{3.30}
$$

with $u_k = \begin{bmatrix} u_k^\alpha, & u_k^\beta \end{bmatrix}^\top$ and

$$f_1\left(x_k^1\right) = \begin{bmatrix} w_{11,k}S\left(\omega_k\right) - \omega_{r,k+1} \\ f_{11,k} \end{bmatrix},$$

$$f_{11,k} = w_{21,k}^2 S^2\left(\omega_k\right)S^2\left(\psi_k^\beta\right) + w_{31,k}^2 S^2\left(\omega_k\right)S^2\left(\psi_k^\alpha\right)$$
$$+ w^2 I_{m,k}^2 - \Psi_{r,k+1},$$

$$I_{m,k} = \sqrt{w_{22,k}^2 i_k^{\alpha 2} + w_{32,k}^2 i_k^{\beta 2}},$$

$$B_1\left(x_k^1\right) = \begin{bmatrix} b_{11,k} & b_{12,k} \\ b_{21,k} & b_{22,k} \end{bmatrix},$$

$$b_{11,k} = w_{12,k}S\left(\omega\right)S\left(\psi_k^\beta\right),$$

$$b_{12,k} = w_{13,k}S\left(\omega\right)S\left(\psi^\alpha\right),$$

$$b_{21,k} = 2w_{31,k}w_{32,k}S\left(\omega_k\right)S\left(\psi_k^\alpha\right),$$

$$b_{22,k} = 2w_{21,k}w_{22,k}S\left(\omega_k\right)S\left(\psi_k^\beta\right),$$

$$f_2\left(x_k^2\right) = \begin{bmatrix} f_{21,k} \\ f_{22,k} \end{bmatrix}, \quad B_{2,k} = \begin{bmatrix} w_{44,k} & 0 \\ 0 & w_{54,k} \end{bmatrix},$$

$$f_{21,k} = w_{41,k}S\left(\psi_k^\alpha\right) + w_{42,k}S\left(\psi_k^\beta\right) + w_{43,k}S\left(i_k^\alpha\right),$$

$$f_{22,k} = w_{51,k}S\left(\psi_k^\alpha\right) + w_{52,k}S\left(\psi_k^\beta\right) + w_{53,k}S\left(i_k^\beta\right),$$

Applying the block control technique, we define the following vector $\mathbf{z}_{1,k} = x_k^1$. Then

$$\mathbf{z}_{1,k+1} = f_1\left(x_k^1\right) + B_1\left(x_k^1\right)x_k^2 = \mathbf{K}\mathbf{z}_{1,k}, \tag{3.31}$$

where $\mathbf{K} = diag\{\mathbf{k}_1, \mathbf{k}_2\}$, with $|\mathbf{k}_i| < 1$ ($i = 1, 2$); then the desired value x_k^{2d} of x_k^2 is calculated from (3.31) as

$$x_k^{2d} = B_1^{-1}\left(x_k^1\right)\left[-f_1\left(x_k^1\right) + \mathbf{K}\mathbf{z}_{1,k}\right].$$

It is desired that $x_k^2 = x_k^{2d}$. For that, a second new error vector is designed:

$$\mathbf{z}_{2,k} = x_k^2 - x_k^{2d}.$$

Then

$$\mathbf{z}_{2,k+1} = f_3\left(x_k^1\right) + B_{2,k}u_k,$$

with

$$f_3\left(x_k^1\right) = f_2\left(x_k^2\right) - B_1^{-1}\left(x_{k+1}^1\right)\left[-f_1\left(x_{k+1}^1\right) + \mathbf{K}\mathbf{z}_{1,k+1}\right].$$

Let us select the manifold for the sliding mode as $S_{D,k} = \mathbf{z}_{2,k}$. In order to design a control law, a discrete-time sliding mode version is implemented as

$$u_k = \begin{cases} u_{eq,k} & \text{if } \left\|u_{eq,k}\right\| \leq u_0 \\ u_0 \frac{u_{eq,k}}{\left\|u_{eq,k}\right\|} & \text{if } \left\|u_{eq,k}\right\| > u_0 \end{cases},$$

where $u_{eq,k} = -B_{2,k}^{-1}f_3\left(x_k^1\right)$ is calculated from $S_{D,k} = 0$ and u_0 are the control resources that bound the control. This system represents the sliding mode dynamics which achieves the control objectives. It is an obvious fact that the proposed control u_k depends on $i_k^{\alpha^2}$ and $i_k^{\beta^2}$, which appears in $f_1(\bullet)$, making the system insolvable [2]. To overcome this problem an observer only is designed with current measurements for the new variable $I_{m,k}$ [2]. Due to the time varying RHONN weights, we need to guarantee that $B_1(\bullet)$ and $B_2(\bullet)$ are not singular; then it is necessary to avoid the zero-crossing of the weights $w_{13,k}$, $w_{22,k}$, $w_{32,k}$, $w_{44,k}$ and $w_{54,k}$, which are the so-called controllability weights [1]. It is important to remark that in this application only the weights $w_{44,k}$ and $w_{54,k}$ could cross zero.

3.4.3 REDUCED ORDER NONLINEAR OBSERVER

The last control algorithm works with the full state measurement assumption [2]. However, the rotor fluxes measurement is a difficult task. Here, a reduced order nonlinear observer is designed for fluxes with rotor speed and current measurements only. The flux dynamics in (3.19) can be written as

$$\Psi(k+1) = aG(k)\Psi(k) + (1-a)MG(k)I(k)$$

with

$$G(k) = \begin{bmatrix} \cos(n_P T \omega(k)) & -\sin(n_P T \omega(k)) \\ \sin(n_P T \omega(k)) & \cos(n_P T \omega(k)) \end{bmatrix},$$

$$I(k) = \begin{bmatrix} i^\alpha(k) \\ i^\beta(k) \end{bmatrix}. \tag{3.32}$$

The proposed observer for system (3.19) assumes the speed and current available for measurements:

$$\hat{\Psi}(k+1) = aG(k)\hat{\Psi}(k) + (1-a)MG(k)I(k).$$

Let us define

$$e^\Psi(k) = \Psi(k) - \hat{\Psi}(k).$$

Then

$$e^\Psi(k+1) = aG(k)e^\Psi(k).$$

A Lyapunov candidate function to prove stability of $e^\Psi(k)$ is

$$V(k) = e^{\Psi^T}(k)e^\Psi(k), \tag{3.33}$$

with

$$\begin{aligned} \Delta V(k) &= V(k+1) - V(k) \\ &= e^{\Psi^T}(k+1)e^\Psi(k+1) - e^{\Psi^T}(k)e^\Psi(k) \\ &= e^{\Psi^T}(k)(a^2 G^T(k)G(k) - I)e^\Psi(k), \end{aligned}$$

TABLE 3.1

Induction Motor Parameters

Parameter	Value	Description
R_s	14Ω	Stator resistance
L_s	$400mH$	Stator inductance
M	$377mH$	Mutual inductance
R_r	10.1Ω	Rotor resistance
L_r	$412.8mH$	Rotor inductance
n_p	2	Number of pole pairs
J	$0.01Kgm^2$	Moment of inertia
ω_n	$168.5rad/s$	Nominal speed
T_{L_n}	$1.1Nm$	Nominal load
T	$0.0001s$	Sampling period

where

$$a^2 G^T(k)G(k) - I < 0. \tag{3.34}$$

By (3.33), $G^T(k)G(k) = I$, then condition (3.34) is reduced to

$$\begin{bmatrix} a^2 & 0 \\ 0 & a^2 \end{bmatrix} - \begin{bmatrix} 1 & 0 \\ 0 & 1 \end{bmatrix} < 0,$$

where $a < 1$, $a = e^{-\alpha T}$. This condition is satisfied due to the fact that T and α are always positive. So the increment of the Lyapunov function (3.33) is always negative, implying that the tracking error tends asymptotically to zero. Now we use $\hat{\Psi}^\alpha$ and $\hat{\Psi}^\beta$ to implement the control algorithm developed above.

3.4.4 SIMULATION RESULTS

Simulations are performed for system (3.28), using parameters as in Table 3.1.

For simulations, the full state measurement assumption is necessary [2]. However, rotor fluxe measurement is a difficult task. Here, the reduced order nonlinear observer described above is used to perform the simulation. The tracking results are presented in Figures 3.7 and 3.8. There, the tracking and identification performance can be verified for the two plant outputs. Figure 3.9 displays the load torque applied as an external disturbance. Figure 3.10 presents the parametric variation introduced in the rotor resistance (R_r) as a variation of 1 ohm per second. Figure 3.11 shows the weights evolution. Figures 3.12 and 3.13 portray the fluxes and their estimates.

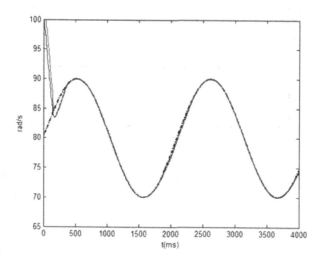

FIGURE 3.7   Tracking performance ω_k (solid line), $x_{1,k}$ (dash-dot line) and $\omega_{r,k}$ (dashed line).

3.5   CONCLUSIONS

This chapter has presented the application of recurrent high order neural networks to design a block control algorithm for a class of discrete-time nonlinear systems. The RHONN is used to perform system identification; the training of the neural networks is performed on-line using an extended Kalman filter in a series-parallel configuration. The boundedness of the identification error is established on the basis of the Lyapunov approach. The proposed training algorithm avoids singularities in the control law due to weight zero-crossings. Simulation results illustrate the robustness of the proposed control methodology, with respect to external disturbances as well as parametric variations.

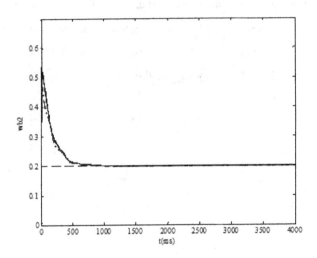

FIGURE 3.8   Tracking performance Ψ_k (solid line), $x_2^2 + x_3^2$ (dash-dot line) and $\Psi_{d,k}$ (dashed line).

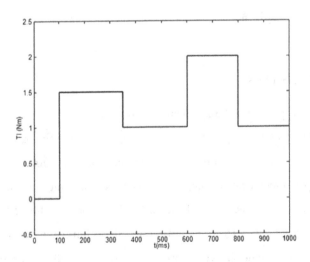

FIGURE 3.9   Load torque $(T_{L,k})$.

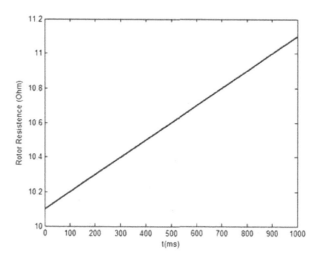

FIGURE 3.10   Rotor resistance variation (R_r).

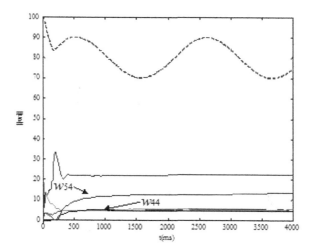

FIGURE 3.11   Weights evolution.

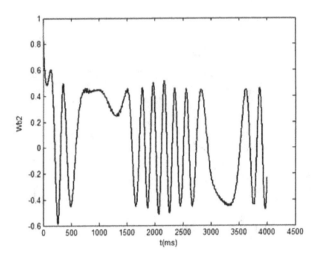

FIGURE 3.12   Time evolution of ψ_k^α and its estimate (real in solid line and estimated in dashed line).

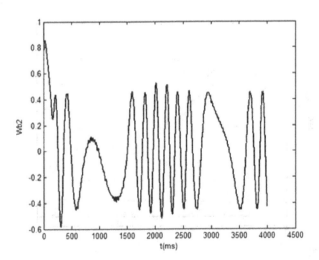

FIGURE 3.13   Time evolution of ψ_k^β and its estimate (real in solid line and estimated in dashed line).

REFERENCES

1. R. A. Felix, E. N. Sanchez and A.G. Loukianov. Avoiding controller singularities in adaptive recurrent neural control. *Proceedings IFAC'05*, Prague, Czech Republic, July, 2005.

2. A. G. Loukianov, J. Rivera and J. M. Cañedo. Discrete-time sliding mode control of an induction motor. *Proceedings IFAC'02*, Barcelone, Spain, July, 2002.

3. G. A. Rovithakis and M. A. Chistodoulou. *Adaptive Control with Recurrent High-Order Neural Networks*. Springer Verlag, Berlin, Germany, 2000.

4. Y. Song and J. W. Grizzle. The extended Kalman filter as local asymptotic observer for discrete-time nonlinear systems. *Journal of Mathematical Systems, Estimation and Control*, 5(1):59–78, 1995.

5. V. Utkin, J. Guldner and J. Shi. *Sliding Mode Control in Electromechanical Systems*. Taylor and Francis, Philadelphia, USA, 1999.

6. W. Yu and X. Li. Nonlinear system identification using discrete-time recurrent neural networks with stable learning algorithms. *Information Sciences*, 158: 131–147, 2004.

4 Neural Optimal Control

This chapter deals with neural optimal control for both stabilization and trajectory tracking. First, the inverse optimal control is established as well as its solution by proposing a quadratic CLF. Next, a robust inverse optimal control is proposed for a disturbed nonlinear system. After that, the inverse optimal control technique for positive systems based on the CLF approach is developed, followed by the inverse optimal control based on a CLF combined with a discrete-time speed-gradient algorithm. Then, the neural inverse optimal control is proposed, achieving stabilization and trajectory tracking for uncertain nonlinear systems, by using a RHONN identifier. The training of the neural network is performed on-line using an extended Kalman filter. A block transformation of the neural model to solve the inverse optimal trajectory tracking as a stabilization problem for block-control-form nonlinear systems is included. Applicability of the proposed control schemes is illustrated via simulations.

4.1 INVERSE OPTIMAL CONTROL VIA CLF

Due to favorable stability margins of optimal control systems, we synthesize a stabilizing feedback control law, which will be optimal with respect to a cost functional. At the same time, we want to avoid the difficult task of solving the associated HJB partial differential equation. In the inverse optimal control approach, a CLF candidate is used to construct an optimal control law directly without solving the HJB equation [7], and still allowing us to obtain Kalman-type stability margins [14]. A storage function is used as a CLF candidate and the inverse optimal control law is selected as an output feedback one, which is obtained as a result of solving the Bellman equation. Then, a CLF candidate for the obtained control law is proposed such that it stabilizes the system and *a posteriori* cost functional is minimized. For this control scheme, a quadratic CLF candidate is used to synthesize the inverse optimal control law. We establish the following assumption and definition which allow the inverse optimal control solution via the CLF approach.

Assumption 4.1 The full state of system (2.1) is measurable.

Along the lines of [31], we propose the discrete-time inverse optimal control law for nonlinear systems as follows.

Definition 4.1: Inverse Optimal Control Law The control law

$$u_k^* = -\frac{1}{2}R^{-1}g^T(x_k)\frac{\partial V(x_{k+1})}{\partial x_{k+1}} \tag{4.1}$$

is inverse optimal if

(i)   it achieves (global) exponential stability of the equilibrium point $x_k = 0$ for system (2.1);

(ii)   it minimizes a cost functional defined as (2.2), for which $l(x_k) := -\overline{V}$ with

$$\overline{V} := V(x_{k+1}) - V(x_k) + u_k^{*T}Ru_k^* \le 0. \tag{4.2}$$

As established in Definition 4.1, inverse optimal control is based on knowledge of $V(x_k)$; thus, we propose a CLF based on $V(x_k)$ such that (i) and (ii) can be guaranteed. That is, instead of solving (2.10) for $V(x_k)$, we propose a control Lyapunov function $V(x_k)$ as

$$V(x_k) = \frac{1}{2}x_k^T P x_k, \qquad P = P^T > 0 \tag{4.3}$$

for control law (4.1) in order to ensure stability of system (2.1) equilibrium point $x_k = 0$, which will be achieved by defining an appropriate matrix P. Moreover, it will be established that control law (4.1) with (4.3), which is referred to as the inverse optimal control law, optimizes a cost functional of the form (2.2). Consequently, by considering $V(x_k)$ as in (4.3), control law (4.1) takes the following form:

$$
\begin{aligned}
u_k^* &= -\frac{1}{2}R^{-1}g^T(x_k)\frac{\partial V(x_{k+1})}{\partial x_{k+1}} \\
&= -\frac{1}{2}R^{-1}g^T(x_k)(Px_{k+1}) \\
&= -\frac{1}{2}R^{-1}g^T(x_k)(Pf(x_k)+Pg(x_k)u_k^*).
\end{aligned}
$$

Thus,

$$
\left(I+\frac{1}{2}R^{-1}g^T(x_k)Pg(x_k)\right)u_k^* = \\
-\frac{1}{2}R^{-1}g^T(x_k)Pf(x_k). \tag{4.4}
$$

Multiplying (4.4) by R, then

$$
\left(R+\frac{1}{2}g^T(x_k)Pg(x_k)\right)u_k^* = -\frac{1}{2}g^T(x_k)Pf(x_k), \tag{4.5}
$$

which results in the following state feedback control law:

$$
\begin{aligned}
\alpha(x_k) &:= u_k^* \\
&= -\frac{1}{2}(R+P_2(x_k))^{-1}P_1(x_k), \tag{4.6}
\end{aligned}
$$

where $P_1(x_k) = g^T(x_k)Pf(x_k)$ and $P_2(x_k) = \frac{1}{2}g^T(x_k)Pg(x_k)$. Note that $P_2(x_k)$ is a positive definite and symmetric matrix, which ensures that the inverse matrix in (4.6) exists.

Once we have proposed a CLF for solving the inverse optimal control in accordance with Definition 4.1, the main contribution of this chapter is presented as the following theorem.

Theorem 4.1

Consider the affine discrete-time nonlinear system (2.1). If there exists a matrix $P = P^T > 0$ such that the following inequality holds

$$V_f(x_k) - \frac{1}{4} P_1^T(x_k)(R + P_2(x_k))^{-1}$$
$$\times P_1(x_k) \leq -\zeta_Q \|x_k\|^2, \tag{4.7}$$

where $V_f(x_k) = V(f(x_k)) - V(x_k)$, with $V(f(x_k)) = \frac{1}{2} f^T(x_k) P f(x_k)$ and $\zeta_Q > 0$; $P_1(x_k)$ and $P_2(x_k)$ are as defined in (4.6), then the equilibrium point $x_k = 0$ of system (2.1) is globally exponentially stabilized by the control law (4.6), with CLF (4.3).

Moreover, with (4.3) as a CLF, this control law is inverse optimal in the sense that it minimizes the cost functional given by

$$V(x_k) = \sum_{k=0}^{\infty} \left(l(x_k) + u_k^T R u_k \right), \tag{4.8}$$

with

$$l(x_k) = -\overline{V}\big|_{u_k^* = \alpha(x_k)} \tag{4.9}$$

and optimal value function $V^*(x_0) = V(x_0)$. ∎

Proof

First, we analyze stability. Global stability for the equilibrium point $x_k = 0$ of system (2.1) with (4.6) as input is achieved if (4.2) is satisfied. Thus, $\overline{V}$ results in

$$
\begin{aligned}
\overline{V} &= V(x_{k+1}) - V(x_k) + \alpha^T(x_k) R \alpha(x_k) \\
&= \frac{f^T(x_k) P f(x_k) + 2 f^T(x_k) P g(x_k) \alpha(x_k)}{2} \\
&\quad + \frac{\alpha^T(x_k) g^T(x_k) P g(x_k) \alpha(x_k) - x_k^T P x_k}{2} + \alpha^T(x_k) R \alpha(x_k) \\
&= V_f(x_k) - \frac{1}{2} P_1^T(x_k)(R + P_2(x_k))^{-1} P_1(x_k) \\
&\quad + \frac{1}{4} P_1^T(x_k)(R + P_2(x_k))^{-1} P_1(x_k) \\
&= V_f(x_k) - \frac{1}{4} P_1^T(x_k)(R + P_2(x_k))^{-1} P_1(x_k).
\end{aligned}
\tag{4.10}
$$

Selecting P such that $\overline{V} \le 0$, the stability of $x_k = 0$ is guaranteed. Furthermore, by means of P, we can achieve a desired negativity amount [8] for the closed-loop function $\overline{V}$ in (4.10). This negativity amount can be bounded using a positive definite matrix Q as follows:

$$
\begin{aligned}
\overline{V} &= V_f(x_k) - \frac{1}{4} P_1^T(x_k)(R + P_2(x_k))^{-1} P_1(x_k) \\
&\le -x_k^T Q x_k \\
&\le -\lambda_{min}(Q) \|x_k\|^2 \\
&= -\zeta_Q \|x_k\|^2, \qquad \zeta_Q = \lambda_{min}(Q),
\end{aligned}
\tag{4.11}
$$

where $\|\cdot\|$ stands for the Euclidean norm and $\zeta_Q > 0$ denotes the minimum eigenvalue of matrix Q ($\lambda_{min}(Q)$). Thus, from (4.11) follows condition (4.7). Considering (4.10)–(4.11), $\overline{V} = V(x_{k+1}) - V(x_k) + \alpha^T(x_k) R \alpha(x_k) \le -\zeta_Q \|x_k\|^2 \Rightarrow \Delta V = V(x_{k+1}) - V(x_k) \le -\zeta_Q \|x_k\|^2$. Moreover, as $V(x_k)$ is a radially unbounded function, then the solution $x_k = 0$ of the closed-loop system (2.1) with (4.6) as input is globally exponentially stable according to Theorem 2.2.

When function $-l(x_k)$ is set to be the (4.11) RHS, that is,

$$l(x_k) := -\overline{V}\Big|_{u_k^* = \alpha(x_k)} \tag{4.12}$$

$$= -V_f(x_k) + \frac{1}{4} P_1^T(x_k) (R + P_2(x_k))^{-1} P_1(x_k),$$

then $V(x_k)$ as proposed in (4.3) is a solution of the DT HJB equation (2.10).

In order to obtain the optimal value for the cost functional (4.8), we substitute $l(x_k)$ given in (4.12) into (4.8); then

$$
\begin{aligned}
V(x_k) &= \sum_{k=0}^{\infty} \left(l(x_k) + u_k^T R u_k \right) \\
&= \sum_{k=0}^{\infty} \left(-\overline{V} + u_k^T R u_k \right) \\
&= -\sum_{k=0}^{\infty} \left[V_f(x_k) - \frac{1}{4} P_1^T(x_k) (R + P_2(x_k))^{-1} P_1(x_k) \right] + \sum_{k=0}^{\infty} u_k^T R u_k.
\end{aligned}
\tag{4.13}
$$

Factorizing (4.13), and then adding the identity matrix

$$I_m = (R + P_2(x_k)) (R + P_2(x_k))^{-1},$$

with $I_m \in \mathbb{R}^{m \times m}$, we obtain

$$
\begin{aligned}
V(x_k) = &-\sum_{k=0}^{\infty} \left[V_f(x_k) - \frac{1}{2} P_1^T(x_k) (R + P_2(x_k))^{-1} P_1(x_k) \right. \\
&+ \frac{1}{4} P_1^T(x_k) (R + P_2(x_k))^{-1} P_2(x_k) (R + P_2(x_k))^{-1} P_1(x_k) \\
&+ \frac{1}{4} P_1^T(x_k) (R + P_2(x_k))^{-1} R \\
&\left. \times (R + P_2(x_k))^{-1} P_1(x_k) \right] + \sum_{k=0}^{\infty} u_k^T R u_k.
\end{aligned}
\tag{4.14}
$$

Being $\alpha(x_k) = -\frac{1}{2}(R + P_2(x_k))^{-1} P_1(x_k)$, then (4.14) becomes

$$
\begin{aligned}
V(x_k) &= -\sum_{k=0}^{\infty}\left[V_f(x_k) + P_1^T(x_k)\alpha(x_k) + \alpha^T(x_k)P_2(x_k)\alpha(x_k)\right] + \sum_{k=0}^{\infty}\left[u_k^T R u_k\right. \\
&\quad \left. -\alpha^T(x_k)R\alpha(x_k)\right] \\
&= -\sum_{k=0}^{\infty}\left[V(x_{k+1}) - V(x_k)\right] + \sum_{k=0}^{\infty}\left[u_k^T R u_k - \alpha^T(x_k)R\alpha(x_k)\right],
\end{aligned}
\tag{4.15}
$$

which can be written as

$$
\begin{aligned}
V(x_k) &= -\sum_{k=1}^{\infty}\left[V(x_{k+1}) - V(x_k)\right] - V(x_1) + V(x_0) \\
&\quad + \sum_{k=0}^{\infty}\left[u_k^T R u_k - \alpha^T(x_k)R\alpha(x_k)\right] \\
&= -\sum_{k=2}^{\infty}\left[V(x_{k+1}) - V(x_k)\right] - V(x_2) + V(x_1) \\
&\quad - V(x_1) + V(x_0) + \sum_{k=0}^{\infty}\left[u_k^T R u_k - \alpha^T(x_k)R\alpha(x_k)\right].
\end{aligned}
\tag{4.16}
$$

For notation convenience in (4.16), the upper limit ∞ will be treated as $N \to \infty$, and then

$$
\begin{aligned}
V(x_k) &= -V(x_N) + V(x_{N-1}) - V(x_{N-1}) + V(x_0) \\
&\quad + \lim_{N\to\infty}\sum_{k=0}^{N}\left[u_k^T R u_k - \alpha^T(x_k)R\alpha(x_k)\right] \\
&= -V(x_N) + V(x_0) + \lim_{N\to\infty}\sum_{k=0}^{N}\left[u_k^T R u_k\right. \\
&\quad \left. -\alpha^T(x_k)R\alpha(x_k)\right].
\end{aligned}
$$

Letting $N \to \infty$ and noting that $V(x_N) \to 0$ for all x_0, then

$$
V(x_k) = V(x_0) + \sum_{k=0}^{\infty}\left[u_k^T R u_k - \alpha^T(x_k)R\alpha(x_k)\right].
\tag{4.17}
$$

Thus, the minimum value of (4.17) is reached with $u_k = \alpha(x_k)$. Hence, the control law (4.6) minimizes the cost functional (4.8). The optimal value function of (4.8) is $V^*(x_k) = V(x_0)$ for all x_0.   ∎

Comment 4.1 Additionally, with $l(x_k)$ as defined in (4.9), $V(x_k)$ solves the following Hamilton–Jacobi–Bellman equation:

$$l(x_k) + V(x_{k+1}) - V(x_k) + \frac{1}{4}\frac{\partial V^T(x_{k+1})}{\partial x_{k+1}}g(x_k)$$

$$\times R^{-1}g^T(x_k)\frac{\partial V(x_{k+1})}{\partial x_{k+1}} = 0. \tag{4.18}$$

It can establish the main conceptual differences between optimal control and inverse optimal control as follows:

- For optimal control, the state cost function $l(x_k) \geq 0$ and the input weighing term $R > 0$ are given a priori. Then, they are used to determine $u(x_k)$ and $V(x_k)$ by means of the discrete-time HJB equation solution.
- For inverse optimal control, the control Lyapunov function $V(x_k)$ and the input weighting term R are given a priori. Then, these functions are used to compute $u(x_k)$ and $l(x_k)$ defined as $l(x_k) := -\bar{V}$.

Optimal control will be in general given as (4.1), and the minimum value of the cost functional (4.8) will be a function of the initial state x_0, that is, $V(x_0)$. If system (2.1) and the control law (4.1) are given, we shall say that the pair $\{V(x_k), l(x_k)\}$ is a solution to the *inverse optimal control* if the performance index (2.2) is minimized by (4.1), the minimum value being $V(x_0)$ [23].

4.1.1 EXAMPLE

The applicability of the developed method is illustrated by synthesis of a stabilizing control law for a discrete-time second order nonlinear system (unstable for $u_k = 0$)

of the form (2.1), with

$$f(x_k) = \begin{bmatrix} x_{1,k}x_{2,k} - 0.8x_{2,k} \\ x_{1,k}^2 + 1.8x_{2,k} \end{bmatrix} \qquad (4.19)$$

and

$$g(x_k) = \begin{bmatrix} 0 \\ -2 + \cos(x_{2,k}) \end{bmatrix}. \qquad (4.20)$$

According to (4.6), the stabilizing optimal control law is formulated as

$$\alpha(x_k) = -\frac{1}{2}\left(R + \frac{1}{2}g^T(x_k)Pg(x_k)\right)^{-1}g^T(x_k)Pf(x_k),$$

where the positive definite matrix P is selected as

$$P = \begin{bmatrix} 10 & 0 \\ 0 & 10 \end{bmatrix}$$

and R is selected as the constant term $R = 1$.

The state penalty term $l(x_k)$ in (4.8) is calculated according to (4.9). The phase portrait for this unstable open-loop ($u_k = 0$) system with initial conditions $\chi_0 = [2\,,\,-2]^T$ is displayed in Figure 4.1.

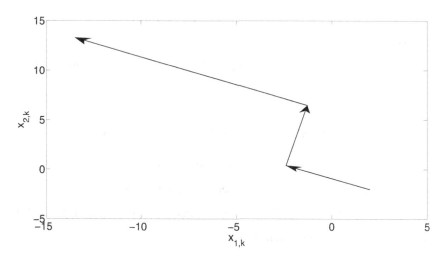

FIGURE 4.1   Unstable phase portrait.

Figure 4.2 presents the time evolution of x_k for this system with initial conditions $x_0 = [2,\ -2]^T$ under the action of the proposed control law. This figure also includes the applied inverse optimal control law, which achieves stability; the respective phase portrait is displayed in Figure 4.3. Figure 4.4 displays the evaluation of the cost

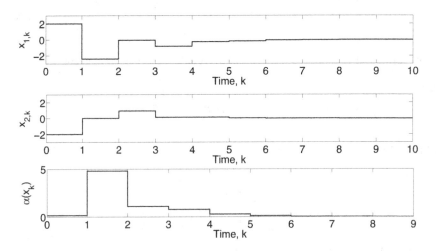

FIGURE 4.2   Stabilization of a nonlinear system.

functional.

Comment 4.2 For this example, according to Theorem 4.1, the optimal value function is calculated as $V^*(x_k) = V(x_0) = \frac{1}{2} x_0^T P x_0 = 40$, which is reached as shown in Figure 4.4.

4.1.2 INVERSE OPTIMAL CONTROL FOR LINEAR SYSTEMS

For the special case of linear systems, it can be established that inverse optimal control is an alternative way to achieve stability and the minimization of a cost functional, avoiding the need to solve the discrete-time algebraic Riccati equation (DARE) [1]. That is, for the discrete-time linear system

$$x_{k+1} = A x_k + B u_k, \qquad x_0 = x(0), \qquad (4.21)$$

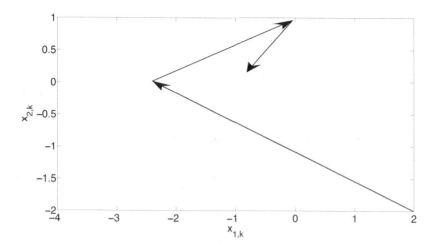

FIGURE 4.3   Phase portrait for the stabilized system.

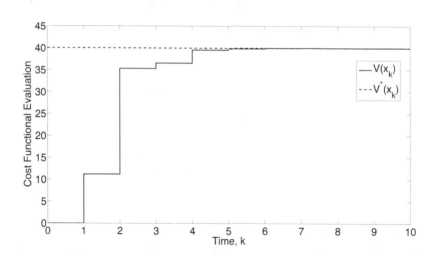

FIGURE 4.4   Cost functional evaluation.

the stabilizing inverse control law (4.6) becomes

$$
\begin{aligned}
u_k^* &= -\frac{1}{2}(R + P_2(x_k))^{-1} P_1(x_k) \\
&= -\frac{1}{2}(R + \frac{1}{2}B^T P B)^{-1} B^T P A x_k,
\end{aligned}
\tag{4.22}
$$

where $P_1(x_k)$ and $P_2(x_k)$, by considering $f(x_k) = A x_k$ and $g(x_k) = B$, are defined as

$$
P_1(x_k) = B^T P A x_k,
\tag{4.23}
$$

and

$$
P_2(x_k) = \frac{1}{2} B^T P B.
\tag{4.24}
$$

Selecting $R = \frac{1}{2}\overline{R} > 0$, where $\overline{R}$ is a constant, (4.22) results in

$$
\begin{aligned}
u_k^* &= -\frac{1}{2}(\frac{1}{2}\overline{R} + \frac{1}{2}B^T P B)^{-1} B^T P A x_k \\
&= -(\overline{R} + B^T P B)^{-1} B^T P A x_k.
\end{aligned}
\tag{4.25}
$$

For this linear case, function $\overline{V}$, as given in (4.10), becomes

$$
\begin{aligned}
\overline{V} &= V_f(x_k) - \frac{1}{4}P_1^T(x_k)(R + P_2(x_k))^{-1} P_1(x_k) \\
&= \frac{x_k^T A^T P A x_k}{2} - \frac{x_k^T P x_k}{2} \\
&\quad \frac{-x_k^T A^T P B(\overline{R} + B^T P B)^{-1} B^T P A x_k}{} \\
&\quad + \frac{u_k^{*T} B^T P B u_k^*}{2} + \frac{u_k^{*T} \overline{R}(x_k) u_k^*}{2}.
\end{aligned}
\tag{4.26}
$$

By means of P, we can achieve a desired negativity amount [8] for the function $\overline{V}$ in (4.26). This negativity amount can be bounded using a positive definite matrix $Q = \frac{1}{2}\overline{Q} > 0$, where $\overline{Q}$ is a constant, as follows:

$$
\overline{V} \le -\frac{x_k^T \overline{Q} x_k}{2}.
\tag{4.27}
$$

If we determine P such that (4.27) is satisfied, then the closed-loop system (4.21) with the inverse optimal control law (4.25) is globally exponentially stable. Note that conditions (4.26)–(4.27) are the linear versions of (4.7). Moreover, inequality (4.27) can be written as

$$
\begin{aligned}
x_k^T P x_k \;=\; & x_k^T \overline{Q} x_k + x_k^T A^T P A x_k \\
& -2 x_k^T A^T P B (\overline{R} + B^T P B)^{-1} B^T P A x_k \\
& + x_k^T A^T P B (\overline{R} + B^T P B)^{-1} \\
& \times (\overline{R} + B^T P B)(\overline{R} + B^T P B)^{-1} B^T P A x_k.
\end{aligned}
\tag{4.28}
$$

Finally, from (4.28) the discrete-time algebraic Riccati equation [1]

$$
P = \overline{Q} + A^T P A - A^T P B (\overline{R} + B^T P B)^{-1} B^T P A
\tag{4.29}
$$

is obtained, as in the case of the linear optimal regulator [1, 3].

The cost functional, which is minimized for the inverse optimal control law (4.25), results in

$$
\begin{aligned}
V(x_k) \;=\; & \sum_{k=0}^{\infty} \left(l(x_k) + \frac{1}{2} u_k^T \overline{R} u_k \right) \\
=\; & \frac{1}{2} \sum_{k=0}^{\infty} \left(u_k^T \overline{Q} u_k + u_k^T \overline{R} u_k \right),
\end{aligned}
\tag{4.30}
$$

where $l(x_k)$ is selected as $l(x_k) := -\overline{V} = \frac{1}{2} x_k^T \overline{Q} x_k$.

4.2 ROBUST INVERSE OPTIMAL CONTROL

Optimal controllers are known to be robust with respect to certain plant parameter variations, disturbances, and unmodeled dynamics as provided by stability margins, which implies that the Lyapunov difference $\Delta V < 0$ might still hold even for internal and/or external disturbances in the plant, and therefore stability will be maintained [21].

In this section, we establish a robust inverse optimal controller achieving disturbance attenuation for a disturbed discrete-time nonlinear system. At the same time,
this controller is optimal with respect to a cost functional, and we avoid solving the
Hamilton–Jacobi–Isaacs (HJI) partial differential equation [7].

Let us consider the disturbed discrete-time nonlinear system

$$x_{k+1} = f(x_k) + g(x_k) u_k + d_k, \qquad x_0 = x(0), \tag{4.31}$$

where $d_k \in \mathbb{R}^n$ is a disturbance which is bounded by

$$\|d_k\| \le \ell_k' + \alpha_4(\|x_k\|), \tag{4.32}$$

with $\ell_k' \le \ell$ a positive constant and $\alpha_4(\|x_k\|)$ a $\mathcal{K}_\infty$- function, and suppose that
$\alpha_4(\|x_k\|)$ in (4.32) is of the same order as the $\mathcal{K}_\infty$- function $\alpha_3(\|x_k\|)$, i.e.,

$$\alpha_4(\|x_k\|) = \delta \, \alpha_3(\|x_k\|), \qquad \delta > 0. \tag{4.33}$$

In the next definition, we establish the discrete-time *robust inverse optimal* control
law.

Definition 4.2: Robust Inverse Optimal Control Law The control law

$$u_k^* = \alpha(x_k) = -\frac{1}{2} R^{-1} g^T(x_k) \frac{\partial V(x_{k+1})}{\partial x_{k+1}} \tag{4.34}$$

is robust inverse optimal if

(i) it achieves (global) ISS for system (4.31);
(ii) $V(x_k)$ is (radially unbounded) positive definite such that the inequality

$$\begin{aligned}
\overline{V}_d(x_k, d_k) : \quad &= \quad V(x_{k+1}) - V(x_k) + u_k^T R u_k \\
&\le \quad -\sigma(x_k) + \ell_d \|d_k\|
\end{aligned} \tag{4.35}$$

is satisfied, where $\sigma(x_k)$ is a positive definite function and ℓ_d is a positive constant. The value of function $\sigma(x_k)$ represents a desired amount of negativity [8] of the closed-loop Lyapunov difference $\overline{V}_d(x_k, d_k)$.

For the robust inverse optimal control solution, let us consider the continuous state feedback control law (4.34), with (4.3) as a CLF candidate, where $P \in \mathbb{R}^{n \times n}$ is assumed to be a positive definite and symmetric matrix. Taking one step ahead for (4.3), then control law (4.34) results in (4.6).

Hence, a robust inverse optimal controller is stated as follows.

Theorem 4.2

Consider a disturbed affine discrete-time nonlinear system (4.31). If there exists a matrix $P = P^T > 0$ such that the following inequality holds

$$V_f(x_k) - \frac{1}{4}P_1^T(x_k)(R + P_2(x_k))^{-1}P_1(x_k) \leq -\zeta\alpha_3(\|x_k\|), \quad \forall\|x_k\| \geq \rho(\|d_k\|), \quad (4.36)$$

with δ in (4.33) satisfying

$$\delta < \frac{\eta}{\ell_d}, \quad (4.37)$$

where function $V_f(x_k) = V(f(x_k)) - V(x_k)$, and with $P_1(x_k) = g^T(x_k)Pf(x_k)$ and $P_2(x_k) = \frac{1}{2}g^T(x_k)Pg(x_k)$, ζ, $\ell_d > 0$, $\eta = (1-\theta)\zeta > 0$, $0 < \theta < 1$, and ρ a $\mathcal{K}_\infty$-function, then the solution of the closed-loop system (4.31) and (4.6) is ISS with the ultimate bound γ (i.e., $\|x_k\| \leq \gamma$, $\forall k \geq k_0 + T$) and (4.3) as an ISS–CLF in (2.23)–(2.24). The ultimate bound γ in (2.17) becomes $\gamma = \alpha_3^{-1}\left(\frac{\ell_d\ell}{\theta_1\zeta}\right)\sqrt{\frac{\lambda_{max}(P)}{\lambda_{min}(P)}}$.

Moreover, with (4.3) as an ISS–CLF, control law (4.6) is inverse optimal in the sense that it minimizes the cost functional given as

$$\mathscr{J} = \sup_{d \in \mathscr{D}}\left\{\lim_{\tau \to \infty}\left[V(x_\tau) + \sum_{k=0}^{\tau}\left(l_d(x_k) + u_k^T R u_k + \ell_d\|d_k\|\right)\right]\right\}, \quad (4.38)$$

where $\mathscr{D}$ is the set of locally bounded functions, and

$$l_d(x_k) := -V_d(x_k, d_k)$$

with $V_d(x_k, d_k)$ a negative definite function.   ∎

Proof

First, we analyze stability of system (4.31) with nonvanishing disturbance d_k. It is worth noting that asymptotic stability of $x = 0$ is not reached anymore [12]; the ISS property for the solution of system (4.31) can only be ensured if stabilizability is assumed. Stability analysis for a disturbed system can be treated by two terms; we propose a Lyapunov difference for the nominal system (i.e., $x_{k+1} = f(x_k) + g(x_k)u_k$), denoted by ΔV, and additionally, a difference for disturbed system (4.31) denoted by ΛV. The Lyapunov difference for the disturbed system is defined as

$$\Delta V_d(x_k, d_k) = V(x_{k+1}) - V(x_k). \tag{4.39}$$

Let us first define the function $V_{nom}(x_{k+1})$ as the $k+1$–step using the Lyapunov function $V(x_k)$ for the nominal system (2.1). Then, adding and subtracting $V_{nom}(x_{k+1})$ in (4.39)

$$\Delta V_d(x_k, d_k) = \underbrace{V(x_{k+1}) - V_{nom}(x_{k+1})}_{\Lambda V :=} + \underbrace{V_{nom}(x_{k+1}) - V(x_k)}_{\Delta V :=}. \tag{4.40}$$

From (4.35) with $\sigma(x_k) = \zeta \alpha_3(\|x_k\|)$, $\zeta > 0$, and the control law (4.6), we obtain

$$\begin{aligned} \Delta V &= V_f(x_k) - \frac{1}{4}P_1^T(x_k)(R + P_2(x_k))^{-1}P_1(x_k) \\ &\leq -\zeta \alpha_3(\|x_k\|) \end{aligned}$$

in (4.40), which is ensured by means of $P = P^T > 0$. On the other hand, since $V(x_k)$ is a C^1 (indeed it is C^2 differentiable) function in x_k for all k, then ΛV satisfies condition (2.19) as

$$
\begin{aligned}
|\Lambda V| &\leq \ell_d \| f(x_k) + g(x_k) u_k(x_k) + d_k - f(x_k) - g(x_k) u_k(x_k) \| \\
&= \ell_d \| d_k \| \\
&\leq \ell_d \ell + \ell_d \alpha_4(\|x_k\|),
\end{aligned}
$$

where ℓ and ℓ_d are positive constants. Hence, the Lyapunov difference $\Delta V_d(x_k, d_k)$ for the disturbed system is determined as

$$
\begin{aligned}
\Delta V_d(x_k, d_k) &= \Lambda V + \Delta V \\
&\leq |\Lambda V| + \Delta V \\
&\leq \ell_d \alpha_4(\|x_k\|) + \ell_d \ell + V_f(x_k) - \frac{1}{4} P_1^T(x_k)(R + P_2(x_k))^{-1} P_1(x_k) \\
&\leq -\zeta \alpha_3(\|x_k\|) + \ell_d \alpha_4(\|x_k\|) + \ell_d \ell \qquad (4.41) \\
&= -\zeta \alpha_3(\|x_k\|) + \theta \zeta \alpha_3(\|x_k\|) - \theta \zeta \alpha_3(\|x_k\|) + \ell_d \alpha_4(\|x_k\|) + \ell_d \ell \\
&= -(1-\theta)\zeta \alpha_3(\|x_k\|) - \theta \zeta \alpha_3(\|x_k\|) + \ell_d \alpha_4(\|x_k\|) + \ell_d \ell \\
&= -(1-\theta)\zeta \alpha_3(\|x_k\|) + \ell_d \alpha_4(\|x_k\|), \qquad \forall \|x_k\| \geq \alpha_3^{-1}\left(\frac{\ell_d \ell}{\theta \zeta}\right),
\end{aligned}
$$

where $0 < \theta < 1$. Using condition (4.33) in the previous expression, we obtain

$$
\begin{aligned}
\Delta V_d(x_k, d_k) &\leq -(1-\theta)\zeta \alpha_3(\|x_k\|) + \ell_d \alpha_4(\|x_k\|) \\
&= -\eta \alpha_3(\|x_k\|) + \ell_d \delta \alpha_3(\|x_k\|) \qquad (4.42) \\
&= -(\eta - \ell_d \delta)\alpha_3(\|x_k\|), \qquad \eta = (1-\theta)\zeta > 0,
\end{aligned}
$$

which is negative definite if condition $\delta < \frac{\eta}{\ell_d}$ (4.37) is satisfied. Therefore, if condition (4.37) holds and considering $V(x_k)$ (4.3) as a radially unbounded ISS–CLF, then by Proposition 2.2, the closed-loop system (4.31) and (4.6) are ISS, which implies BIBS stability and $\mathscr{K}$– asymptotic gain according to Theorem 2.3.

By Definition 2.11 and Comment 2.1, the solution of the closed-loop system

(4.31) and (4.6) is ultimately bounded with $\gamma = \alpha_1^{-1} \circ \alpha_2 \circ \rho$, which results on $\gamma = \alpha_3^{-1} \left(\frac{\ell_d \ell}{\theta \zeta} \right) \sqrt{\frac{\lambda_{max}(P)}{\lambda_{min}(P)}}$. Hence, according to Definition 2.6, the solution is ultimately bounded with ultimate bound $b = \gamma$.

In order to establish inverse optimality, considering that the control (4.6) achieves ISS for the system (4.31), and substituting $l_d(x_k)$ in (4.38), it follows that

$$
\begin{aligned}
\mathscr{J} &= \sup_{d \in \mathscr{D}} \left\{ \lim_{\tau \to \infty} \left[V(x_\tau) + \sum_{k=0}^{\tau} \left(l_d(x_k) + u_k^T R u_k + \ell_d \|d_k\| \right) \right] \right\} \\
&= \sup_{d \in \mathscr{D}} \left\{ \lim_{\tau \to \infty} \left[V(x_\tau) + \sum_{k=0}^{\tau} \left(-\Lambda V - \Delta V + u_k^T R u_k + \ell_d \|d_k\| \right) \right] \right\} \\
&= \sup_{d \in \mathscr{D}} \left\{ \lim_{\tau \to \infty} \left[V(x_\tau) - \sum_{k=0}^{\tau} \left(V_f(x_k) - \frac{1}{4} P_1^T(x_k) (R + P_2(x_k))^{-1} P_1(x_k) \right. \right. \right. \\
&\qquad \left. \left. \left. + \ell_d \ell + \ell_d \delta \alpha_3(\|x_k\|) \right) + \sum_{k=0}^{\tau} u_k^T R u_k + \sum_{k=0}^{\tau} \ell_d \|d_k\| \right] \right\} \\
&= \lim_{\tau \to \infty} \left[V(x_\tau) - \sum_{k=0}^{\tau} \left(V_f(x_k) - \frac{1}{4} P_1^T(x_k) (R + P_2(x_k))^{-1} P_1(x_k) \right) \right. \\
&\qquad \left. + \sum_{k=0}^{\tau} u_k^T R u_k + \sup_{d \in \mathscr{D}} \left\{ \sum_{k=0}^{\tau} \left(\ell_d \|d_k\| - \ell_d \ell - \ell_d \delta \alpha_3(\|x_k\|) \right) \right\} \right]. \quad (4.43)
\end{aligned}
$$

Adding the term

$$
\frac{1}{4} P_1^T(x_k) (R + P_2(x_k))^{-1} R (R + P_2(x_k))^{-1} P_1(x_k)
$$

at the first addition term of (4.43) and subtracting at the second addition term of (4.43) yields

$$
\begin{aligned}
\mathscr{J} &= \lim_{\tau \to \infty} \left[V(x_\tau) - \sum_{k=0}^{\tau} (V(x_{k+1}) - V(x_k)) + \sum_{k=0}^{\tau} \left(u_k^T R u_k - \frac{1}{4} P_1^T(x_k) (R \right. \right. \\
&\qquad \left. + P_2(x_k))^{-1} R (R + P_2(x_k))^{-1} P_1(x_k) \right) \\
&\qquad \left. + \sup_{d \in \mathscr{D}} \left\{ \sum_{k=0}^{\tau} \left(\ell_d \|d_k\| - \ell_d \ell - \ell_d \delta \alpha_3(\|x_k\|) \right) \right\} \right]
\end{aligned}
$$

$$
\begin{aligned}
= \; & \lim_{\tau\to\infty}\Bigg[V(x_\tau) - \sum_{k=0}^{\tau}(V(x_{k+1}) - V(x_k)) + \sum_{k=0}^{\tau}\left[u_k^T R u_k - \alpha^T(x_k)R\alpha(x_k)\right] \\
& + \sup_{d\in\mathscr{D}}\left\{ \sum_{k=0}^{\tau}(\ell_d\,\|d_k\| - \ell_d\ell - \ell_d\delta\alpha_3(\|x_k\|))\right\}\Bigg] \\
= \; & \lim_{\tau\to\infty}\Bigg[V(x_\tau) - V(x_\tau) + V(x_0) + \sum_{k=0}^{\tau}\left[u_k^T R u_k - \alpha^T(x_k)R\alpha(x_k)\right] \\
& + \sum_{k=0}^{\tau}\left(\sup_{d\in\mathscr{D}}\{\ell_d\,\|d_k\|\} - \ell_d\ell - \ell_d\delta\alpha_3(\|x_k\|)\right)\Bigg].
\end{aligned} \tag{4.44}
$$

If $\sup_{d\in\mathscr{D}}\{\ell_d\,\|d_k\|\}$ is taken as the worst case by considering the equality for (4.32), we obtain

$$
\begin{aligned}
\sup_{d\in\mathscr{D}}\{\ell_d\,\|d_k\|\} &= \ell_d\sup_{d\in\mathscr{D}}\{\|d_k\|\} \\
&= \ell_d\ell + \ell_d\,\delta\alpha_3(\|x_k\|).
\end{aligned} \tag{4.45}
$$

Therefore

$$
\sum_{k=0}^{\tau}\left(\sup_{d\in\mathscr{D}}\{\ell_d\,\|d_k\|\} - \ell_d\ell - \ell_d\delta\alpha_3(\|x_k\|)\right) = 0. \tag{4.46}
$$

Thus, the minimum value of (4.44) is reached with $u_k = \alpha(x_k)$. Hence, the control law (4.6) minimizes the cost functional (4.38). The optimal value function of (4.38) is $\mathscr{J}^*(x_0) = V(x_0)$. ∎

Comment 4.3 It is worth noting that, in the inverse optimality analysis proof, equality for (4.33) is considered in order to optimize with respect to the worst case for the disturbance.

As a special case, the manipulation of class $\mathscr{K}_\infty-$ functions in Definition 2.10 is simplified when the class $\mathscr{K}_\infty-$ functions take the special form $\alpha_i(r) = \kappa_i r^c$, $\kappa_i > 0$, $c > 1$, and $i = 1,2,3$. In this case, exponential stability is achieved [12]. Let us assume that the disturbance term d_k in (4.31) satisfies the bound

$$\|d_k\| \le \ell + \delta \|x_k\|^2, \tag{4.47}$$

where ℓ and δ are positive constants, then the following result is stated.

Corollary 4.1

Consider the disturbed affine discrete-time nonlinear system (4.31) with (4.47). If there exists a matrix $P = P^T > 0$ such that the following inequality holds

$$V_f(x_k) - \frac{1}{4} P_1^T(x_k)(R + P_2(x_k))^{-1} P_1(x_k) \le -\zeta_Q \|x_k\|^2 \quad \forall \|x_k\| \ge \rho(\|d_k\|), \tag{4.48}$$

where $\zeta_Q > 0$ denotes the minimum eigenvalue of matrix Q as established in (4.11), and δ in (4.47) satisfies

$$\delta < \frac{\zeta_Q}{\ell_d}, \tag{4.49}$$

then the solution of closed-loop system (4.31), (4.6) is ISS, with (4.3) as an ISS–CLF. The ultimate bound γ in (2.17) becomes $\gamma = \sqrt{\frac{\ell_d \ell}{\theta \eta}} \sqrt{\frac{\lambda_{max}(P)}{\lambda_{min}(P)}}$ with $0 < \theta < 1$ and $\eta > 0$. This bound is reached exponentially.

Moreover, with (4.3) as an ISS–CLF, this control law is inverse optimal in the sense that it minimizes the cost functional given as

$$\mathcal{J} = \sup_{d \in \mathcal{D}} \left\{ \lim_{\tau \to \infty} \left[V(x_\tau) + \sum_{k=0}^{\tau} \left(l_d(x_k) + u_k^T R u_k + \ell_d \|d_k\| \right) \right] \right\} \tag{4.50}$$

where $\mathcal{D}$ is the set of locally bounded functions, and

$$l_d(x_k) = -\Lambda V - \Delta V.$$

Proof

Stability is analyzed similarly to the proof of Theorem 4.2, where the Lyapunov difference is treated by means of two terms as in (4.40). For the first one, the disturbance term is considered, and for the second one, the Lyapunov difference is analyzed in order to achieve exponential stability for an undisturbed system. For the latter, Lyapunov difference ΔV is considered from (4.11); hence the Lyapunov difference ΔV becomes $\Delta V \leq -\zeta_Q \|x_k\|^2$ with a positive constant ζ_Q, and since $V(x_k)$ is a C^1 function in x_k for all k, then ΔV satisfies the bound condition (2.19) as

$$
\begin{aligned}
|\Delta V| \;\; &\leq \;\; \ell_d \|f(x_k) + g(x_k)u_k + d_k - f(x_k) + g(x_k)u_k\| \\
&= \;\; \ell_d \|d_k\| \\
&\leq \;\; \ell_d \ell + \ell_d \delta \|x_k\|^2 ,
\end{aligned}
\tag{4.51}
$$

where ℓ_d and δ are positive constants, and the bound disturbance (4.47) is regarded. Hence, from (4.11) and (4.51) the Lyapunov difference $\Delta V_d(x_k, d_k)$ for disturbed system (4.31) is established as

$$
\begin{aligned}
\Delta V_d(x_k, d_k) \;\; &= \;\; \Delta V + \Delta V \\
&\leq \;\; |\Delta V| + \Delta V \\
&\leq \;\; \ell_d \delta \|x_k\|^2 + \ell_d \ell + V_f(x_k) - \frac{1}{4} P_1^T(x_k)(R + P_2(x_k))^{-1} P_1(x_k) \\
&\leq \;\; -x_k^T Q x_k + \ell_d \delta \|x_k\|^2 + \ell_d \ell \\
&\leq \;\; -\zeta_Q \|x_k\|^2 + \ell_d \delta \|x_k\|^2 + \ell_d \ell \\
&= \;\; -(\zeta_Q - \ell_d \delta) \|x_k\|^2 + \ell_d \ell \\
&= \;\; -\eta \|x_k\|^2 + \ell_d \ell, && \eta = \zeta_Q - \ell_d \delta > 0 \\
&= \;\; -\eta \|x_k\|^2 + \theta \eta \|x_k\|^2 - \theta \eta \|x_k\|^2 + \ell_d \ell, && 0 < \theta < 1 \\
&= \;\; -(1 - \theta)\eta \|x_k\|^2 , && \|x_k\| > \sqrt{\frac{\ell_d \ell}{\theta \eta}} . \tag{4.52}
\end{aligned}
$$

At this point, it must be ensured that η in (4.52) is positive, and thus $\delta < \frac{\zeta_Q}{\ell_d}$, i.e., the condition (4.49).

To this end, as $V(x_k)$ is a radially unbounded function ISS–CLF, then, by Proposition 2.2, the solution of the closed-loop system (4.6), (4.31) is ISS with exponential convergence to the ultimate bound γ, which results in $\gamma = \sqrt{\frac{\ell_d \ell}{\theta \eta}} \sqrt{\frac{\lambda_{max}(P)}{\lambda_{min}(P)}}$.

In order to establish inverse optimality, considering that (4.6) achieves ISS for (4.31), and substituting $l_d(x_k)$ in (4.50), it follows that

$$
\begin{aligned}
\mathscr{J} &= \sup_{d \in \mathscr{D}} \left\{ \lim_{\tau \to \infty} \left[V(x_\tau) + \sum_{k=0}^{\tau} \left(l_d(x_k) + u_k^T R u_k + \ell_d \|d_k\| \right) \right] \right\} \\
&= \sup_{d \in \mathscr{D}} \left\{ \lim_{\tau \to \infty} \left[V(x_\tau) + \sum_{k=0}^{\tau} \left(-\Lambda V - \Delta V + u_k^T R u_k + \ell_d \|d_k\| \right) \right] \right\} \\
&= \sup_{d \in \mathscr{D}} \left\{ \lim_{\tau \to \infty} \left[V(x_\tau) - \sum_{k=0}^{\tau} \left(V_f(x_k) - \frac{1}{4} P_1^T(x_k)(R + P_2(x_k))^{-1} P_1(x_k) \right. \right. \right. \\
&\quad \left. \left. \left. + \ell_d \ell + \ell_d \delta \|x_k\|^2 \right) + \sum_{k=0}^{\tau} u_k^T R u_k + \sum_{k=0}^{\tau} \ell_d \|d_k\| \right] \right\} \\
&= \lim_{\tau \to \infty} \left[V(x_\tau) - \sum_{k=0}^{\tau} \left(V_f(x_k) - \frac{1}{4} P_1^T(x_k)(R + P_2(x_k))^{-1} P_1(x_k) \right) \right. \\
&\quad \left. + \sum_{k=0}^{\tau} u_k^T R u_k + \sup_{d \in \mathscr{D}} \left\{ \sum_{k=0}^{\tau} \left(\ell_d \|d_k\| - \ell_d \ell - \ell_d \delta \|x_k\|^2 \right) \right\} \right].
\end{aligned}
$$

Adding the term $\frac{1}{4} P_1^T(x_k)(R + P_2(x_k))^{-1} R (R + P_2(x_k))^{-1} P_1(x_k)$ to the first addition term and subtracting in the second addition term yields

$$
\begin{aligned}
\mathscr{J} &= \lim_{\tau \to \infty} \left[V(x_\tau) - \sum_{k=0}^{\tau} (V(x_{k+1}) - V(x_k)) + \sum_{k=0}^{\tau} \left(u_k^T R u_k \right. \right. \\
&\quad \left. -\frac{1}{4} P_1^T(x_k)(R + P_2(x_k))^{-1} R (R + P_2(x_k))^{-1} P_1(x_k) \right) \\
&\quad \left. + \sup_{d \in \mathscr{D}} \left\{ \sum_{k=0}^{\tau} \left(\ell_d \|d_k\| - \ell_d \ell - \ell_d \delta \|x_k\|^2 \right) \right\} \right] \\
&= \lim_{\tau \to \infty} \left[V(x_\tau) - \sum_{k=0}^{\tau} (V(x_{k+1}) - V(x_k)) + \sum_{k=0}^{\tau} \left[u_k^T R u_k \right. \right. \\
&\quad \left. \left. -\alpha^T(x_k) R \alpha(x_k) \right] + \sup_{d \in \mathscr{D}} \left\{ \sum_{k=0}^{\tau} \left(\ell_d \|d_k\| - \ell_d \ell - \ell_d \delta \|x_k\|^2 \right) \right\} \right] \\
&= \lim_{\tau \to \infty} \left[V(x_\tau) - V(x_\tau) + V(x_0) + \sum_{k=0}^{\tau} \left[u_k^T R u_k - \alpha^T(x_k) R \alpha(x_k) \right] \right. \\
&\quad \left. + \sum_{k=0}^{\tau} \left(\sup_{d \in \mathscr{D}} \{\ell_d \|d_k\|\} - \ell_d \ell - \ell_d \delta \|x_k\|^2 \right) \right].
\end{aligned}
\tag{4.53}
$$

If $\sup_{d \in \mathscr{D}} \{\ell_d \|d_k\|\}$ is taken as the worst case by considering the equality for (4.47), we obtain

$$
\begin{aligned}
\sup_{d \in \mathscr{D}} \{\ell_d \|d_k\|\} &= \ell_d \sup_{d \in \mathscr{D}} \{\|d_k\|\} \\
&= \ell_d \ell + \ell_d \delta \|x_k\|^2.
\end{aligned}
\tag{4.54}
$$

Therefore

$$
\sum_{k=0}^{\tau} \left(\sup_{d \in \mathscr{D}} \{\ell_d \|d_k\|\} - \ell_d \ell - \ell_d \delta \|x_k\|^2 \right) = 0.
\tag{4.55}
$$

Thus, the minimum value of (4.53) is reached with $u_k = \alpha(x_k)$, and the control law (4.6) minimizes the cost functional (4.50). The optimal value function of (4.50) is $\mathscr{J}^*(x_0) = V(x_0)$.   ∎

Comment 4.4 The terminal penalty $V(x_\tau)$ in (4.38) and (4.50) is a necessary function to avoid imposing the assumption $x_\tau \to 0$ as $\tau \to \infty$. Hence, inverse optimality is guaranteed only outside the ball, which is bounded by function γ as defined in (2.17).

4.3 TRAJECTORY TRACKING INVERSE OPTIMAL CONTROL

Consider the affine discrete-time nonlinear system (2.1). The following cost functional is associated with trajectory tracking for system (2.1):

$$\mathscr{J}(z_k) = \sum_{n=k}^{\infty} \left(l(z_n) + u_n^T R u_n \right), \tag{4.56}$$

where $z_k = x_k - x_{\delta,k}$ with $x_{\delta,k}$ as the desired trajectory for x_k; $z_k \in \mathbb{R}^n$; $\mathscr{J}(z_k) : \mathbb{R}^n \to \mathbb{R}^+$; $l(z_k) : \mathbb{R}^n \to \mathbb{R}^+$ is a positive semidefinite function; and $R : \mathbb{R}^n \to \mathbb{R}^{m \times m}$ is a real symmetric positive definite weighting matrix. The cost functional (4.56) is a performance measure [13]. The entries of R can be fixed or be functions of the system state in order to vary the weighting on control efforts according to the state value [13]. Considering state feedback control, we assume that the full state x_k is available.

Using the optimal value function $\mathscr{J}^*(x_k)$ for (4.56) as Lyapunov function $V(x_k)$, Equation (4.56) can be rewritten as

$$
\begin{aligned}
V(z_k) &= l(z_k) + u_k^T R u_k + \sum_{n=k+1}^{\infty} l(z_n) + u_n^T R u_n \\
&= l(z_k) + u_k^T R u_k + V(z_{k+1}),
\end{aligned}
\tag{4.57}
$$

where we require the boundary condition $V(0) = 0$ so that $V(z_k)$ becomes a Lyapunov function.

Now, we establish the conditions that the optimal control law must satisfy. We define the discrete-time Hamiltonian $\mathscr{H}(z_k, u_k)$ as

$$\mathscr{H}(z_k, u_k) = l(z_k) + u_k^T R u_k + V(z_{k+1}) - V(z_k). \tag{4.58}$$

A necessary condition that the optimal control law should satisfy is $\frac{\partial \mathscr{H}(z_k, u_k)}{\partial u_k} = 0$, then

$$
\begin{aligned}
0 &= 2Ru_k + \frac{\partial V(z_{k+1})}{\partial u_k} \\
&= 2Ru_k + \frac{\partial z_{k+1}}{\partial u_k}\frac{\partial V(z_{k+1})}{\partial z_{k+1}} \\
&= 2Ru_k + g^T(x_k)\frac{\partial V(z_{k+1})}{\partial z_{k+1}}.
\end{aligned} \qquad (4.59)
$$

Therefore, the optimal control law to achieve trajectory tracking is formulated as

$$
u_k^* = -\frac{1}{2}R^{-1}g^T(x_k)\frac{\partial V(z_{k+1})}{\partial z_{k+1}}, \qquad (4.60)
$$

with the boundary condition $V(0) = 0$. For determining the trajectory tracking optimal control, it is necessary to solve the following HJB equation:

$$
l(z_k) + V(z_{k+1}) - V(z_k) + \frac{1}{4}\frac{\partial V^T(z_{k+1})}{\partial z_{k+1}}g(x_k)R^{-1}g^T(x_k)\frac{\partial V(z_{k+1})}{\partial z_{k+1}} = 0, \quad (4.61)
$$

which is a challenging task. To overcome this problem, we propose using inverse optimal control. The main characteristic of inverse optimal control is that a stabilizing feedback control law is designed first, and then it is established that this law optimizes the cost functional (4.56).

Definition 4.3: Tracking Inverse Optimal Control Law Consider the tracking error as $z_k = x_k - x_{\delta,k}$, $x_{\delta,k}$ being the desired trajectory for x_k. Let us define the control law

$$
u_k^* = -\frac{1}{2}R^{-1}g^T(x_k)\frac{\partial V(z_{k+1})}{\partial z_{k+1}}, \qquad (4.62)
$$

which will be inverse optimal stabilizing along the desired trajectory $x_{\delta,k}$ if

(i)   it achieves (global) asymptotic stability of $z_k = 0$ for system (2.1) along the reference $x_{\delta,k}$;

(ii) $V(z_k)$ is a (radially unbounded) positive definite function such that inequality

$$\overline{V} := V(z_{k+1}) - V(z_k) + u_k^{*T} R u_k^* \leq 0 \qquad (4.63)$$

is satisfied.

When we select $l(z_k) := -\overline{V}$, then $V(z_k)$ is a solution for (4.61), and cost functional (4.56) is minimized.

As established in Definition 4.3, the inverse optimal control law for trajectory tracking is based on knowledge of $V(z_k)$. Then, we propose a CLF, $V(z_k)$, such that (i) and (ii) are guaranteed. Hence, instead of solving (4.61) for $V(z_k)$, a quadratic CLF candidate $V(z_k)$ for (4.62) is proposed with the form

$$V(z_k) = \frac{1}{2} z_k^T P z_k, \qquad P = P^T > 0 \qquad (4.64)$$

in order to ensure stability of the tracking error z_k, where

$$
\begin{aligned}
z_k &= x_k - x_{\delta,k} \\
&= \begin{bmatrix} (x_{1,k} - x_{1\delta,k}) \\ (x_{2,k} - x_{2\delta,k}) \\ \vdots \\ (x_{n,k} - x_{n\delta,k}) \end{bmatrix}.
\end{aligned} \qquad (4.65)
$$

Moreover, it will be established that the control law (4.62) with (4.64), which is referred to as the *inverse optimal* control law, optimizes a cost functional as (4.56).

Consequently, by considering $V(x_k)$ as in (4.64), control law (4.62) takes the following form:

$$
\begin{aligned}
u_k^* &= -\frac{1}{4} R g^T (x_k) \frac{\partial z_{k+1}^T P z_{k+1}}{\partial z_{k+1}} \\
&= -\frac{1}{2} R g^T (x_k) P z_{k+1} \\
&= -\frac{1}{2} \left(R + \frac{1}{2} g^T (x_k) P g(x_k) \right)^{-1} g^T (x_k) P \left(f(x_k) - x_{\delta,k+1} \right). \quad (4.66)
\end{aligned}
$$

Once we have proposed a CLF for solving the inverse optimal control for trajectory tracking in accordance with Definition 4.3, the following theorem is presented.

Theorem 4.3

Consider the affine discrete-time nonlinear system (2.1). If there exists a matrix $P = P^T > 0$ such that the following inequality holds:

$$
\begin{aligned}
&\frac{1}{2} f^T (x_k) P f (x_k) + \frac{1}{2} x_{\delta,k+1}^T P x_{\delta,k+1} - \frac{1}{2} x_k^T P x_k \\
&- \frac{1}{2} x_{\delta,k}^T P x_{\delta,k} - \frac{1}{4} P_1^T \left(x_k, x_{\delta,k} \right) (R + P_2 (x_k))^{-1} P_1 \left(x_k, x_{\delta,k} \right) \\
&\leq -\frac{1}{2} \|P\| \, \|f(x_k)\|^2 - \frac{1}{2} \|P\| \, \|x_{\delta,k+1}\|^2 \\
&\quad - \frac{1}{2} \|P\| \, \|x_k\|^2 - \frac{1}{2} \|P\| \, \|x_{\delta,k}\|^2, \quad (4.67)
\end{aligned}
$$

where $P_1 (x_k, x_{\delta,k})$ and $P_2 (x_k)$ are defined as

$$
P_1 (x_k, x_{\delta,k}) = g^T (x_k) P \left(f (x_k) - x_{\delta,k+1} \right) \quad (4.68)
$$

and

$$
P_2 (x_k) = \frac{1}{2} g^T (x_k) P g (x_k) \quad (4.69)
$$

respectively; then system (2.1) with control law (4.66) guarantees asymptotic trajectory tracking along the desired trajectory $x_{\delta,k}$, where $z_{k+1} = x_{k+1} - x_{\delta,k+1}$. ∎

Proof

System (2.1), with control law (4.66), must satisfy inequality (4.63). Considering one step ahead for z_k, we have

$$
\begin{aligned}
\overline{V} &= \frac{1}{2}z_{k+1}^T P z_{k+1} - \frac{1}{2}z_k^T P z_k + u_k^{*T} R u_k^* \\
&= \frac{1}{2}\left(x_{k+1} - x_{\delta,k+1}\right)^T P\left(x_{k+1} - x_{\delta,k+1}\right) - \frac{1}{2}\left(x_k - x_{\delta,k}\right)^T P\left(x_k - x_{\delta,k}\right) \\
&\quad + u_k^{*T} R u_k^*.
\end{aligned}
\tag{4.70}
$$

Substituting (2.1) in (4.70), then

$$
\begin{aligned}
\overline{V} &= \frac{1}{2}\left(f(x_k) + g(x_k)u_k^* - x_{\delta,k+1}\right)^T P\left(f(x_k) + g(x_k)u_k^* - x_{\delta,k+1}\right) \\
&\quad - \frac{1}{2}\left(x - x_{\delta,k}\right)^T P\left(x_k - x_{\delta,k}\right) + u_k^{*T} R u_k^* \\
&= \frac{1}{2}f^T(x_k) P f(x_k) + \frac{1}{2}u_k^{*T} g^T(x_k) P g(x_k) u_k^* + \frac{1}{2}x_{\delta,k+1}^T P x_{\delta,k+1} \\
&\quad + \frac{1}{2}f^T(x_k) P g(x_k) u_k^* + \frac{1}{2}u_k^{*T} g^T(x_k) P f(x_k) - \frac{1}{2}f^T(x_k) P x_{\delta,k+1} \\
&\quad - \frac{1}{2}x_{\delta,k+1}^T P f^T(x_k) - \frac{1}{2}u_k^T g^T(x_k) P x_{\delta,k+1} - \frac{1}{2}x_{\delta,k+1}^T P g(x_k) u_k^* \\
&\quad - \frac{1}{2}x_k^T P x_k - \frac{1}{2}x_{\delta,k}^T P x_{\delta,k} + \frac{1}{2}x_{\delta,k}^T P x_k + \frac{1}{2}x_k^T P x_{\delta,k} + u_k^{*T} R u_k^*.
\end{aligned}
\tag{4.71}
$$

By simplifying, (4.71) becomes

$$
\begin{aligned}
\overline{V} &= \frac{1}{2}f^T(x_k) P f(x_k) + \frac{1}{2}u_k^{*T} g^T(x_k) P g(x_k) u + \frac{1}{2}x_{\delta,k+1}^T P x_{\delta,k+1} \\
&\quad + f^T(x_k) P g(x_k) u_k^* - f^T(x_k) P x_{\delta,k+1} - x_{\delta,k+1}^T P g(x_k) u_k^* \\
&\quad - \frac{1}{2}x_k^T P x_k - \frac{1}{2}x_{\delta,k}^T P x_{\delta,k} + x_k^T P x_{\delta,k} + u_k^{*T} R u_k^* \\
&= \frac{1}{2}f^T(x_k) P f(x_k) + \frac{1}{2}x_{\delta,k+1}^T P x_{\delta,k+1} + x_{\delta,k+1}^T P g(x_k) u_k \\
&\quad - x_{\delta,k+1}^T P g(x_k) u_k^* - f^T(x_k) P x_{\delta,k+1} + x_k^T P x_{\delta,k} - \frac{1}{2}x_k^T P x_k \\
&\quad - \frac{1}{2}x_{\delta,k}^T P x_{\delta,k} + P_1^T(x_k, x_{\delta,k}) u_k^* + u_k^{*T} P_2(x_k) u_k^* + u_k^{*T} R u_k^*,
\end{aligned}
\tag{4.72}
$$

which after using the control law (4.66), (4.72) results in

$$
\begin{aligned}
\overline{V} \;=\; & \frac{1}{2} f^T (x_k) \, P f (x_k) + \frac{1}{2} x_{\delta,k+1}^T P x_{\delta,k+1} - \frac{1}{2} x_k^T P x_k - \frac{1}{2} x_{\delta,k}^T P x_{\delta,k} \\
& - \frac{1}{4} P_1^T (x_k, x_{\delta,k}) \, (R + P_2 (x_k))^{-1} P_1 (x_k, x_{\delta,k}) \\
& - f^T (x_k) \, P x_{\delta,k+1} + x_k^T P x_{\delta,k}.
\end{aligned}
\tag{4.73}
$$

Analyzing the sixth and seventh RHS terms of (4.73) by using the inequality $X^T Y + Y^T X \leq X^T \Lambda X + Y^T \Lambda^{-1} Y$ proved in [30], which is valid for any vector $X \in \mathbb{R}^{n \times 1}$, then for the sixth RHS term of (4.93), we have

$$
\begin{aligned}
f^T (x_k) P x_{\delta,k+1} \;\leq\; & \frac{1}{2} \left[f^T (x_k) P f(x_k) + (P x_{\delta,k+1})^T P^{-1} (P x_{\delta,k+1}) \right] \\
\leq\; & \frac{1}{2} \left[f^T (x_k) P f(x_k) + x_{\delta,k+1}^T P x_{\delta,k+1} \right] \\
\leq\; & \frac{1}{2} \|P\| \, \|f\|^2 + \frac{1}{2} \|P\| \, \|x_{\delta,k+1}\|^2.
\end{aligned}
\tag{4.74}
$$

The seventh RHS term of (4.73) becomes

$$
\begin{aligned}
x_k^T P x_{\delta,k} \;\leq\; & \frac{1}{2} \left[x_k^T P x_k + (P x_{\delta,k})^T P^{-1} (P x_{\delta,k}) \right] \\
\leq\; & \frac{1}{2} \left[x_k^T P x_k + (x_{\delta,k})^T P (x_{\delta,k}) \right].
\end{aligned}
\tag{4.75}
$$

From (4.75), the following expression holds.

$$
\frac{1}{2} \left[x_k^T P x_k + (x_{\delta,k})^T P^{-1} (x_{\delta,k}) \right] \leq \frac{1}{2} \|P\| \, \|x_k\|^2 + \frac{1}{2} \|P\| \, \|x_{\delta,k}\|.
\tag{4.76}
$$

Substituting (4.74) and (4.76) into (4.73), then

$$
\begin{aligned}
\overline{V} \;=\; & \frac{1}{2} f^T (x_k) \, P f (x_k) + \frac{1}{2} x_{\delta,k+1}^T P x_{\delta,k+1} - \frac{1}{2} x_k^T P x_k \\
& - \frac{1}{2} x_{\delta,k}^T P x_{\delta,k} - \frac{1}{4} P_1^T (x_k, x_{\delta,k}) \, (R + P_2 (x_k))^{-1} P_1 (x_k, x_{\delta,k}) \\
& + \frac{1}{2} \|P\| \, \|f (x_k)\|^2 + \frac{1}{2} \|P\| \, \|x_{\delta,k+1}\|^2 \\
& + \frac{1}{2} \|P\| \, \|x_k\|^2 + \frac{1}{2} \|P\| \, \|x_{\delta,k}\|^2 .
\end{aligned}
\tag{4.77}
$$

In order to achieve asymptotic stability, it is required that $\overline{V} \leq 0$, then from (4.77) the next inequality is formulated

$$
\begin{aligned}
& \frac{1}{2} f^T (x_k) \, P f (x_k) + \frac{1}{2} x_{\delta,k+1}^T P x_{\delta,k+1} - \frac{1}{2} x_k^T P x_k \\
& - \frac{1}{2} x_{\delta,k}^T P x_{\delta,k} - \frac{1}{4} P_1^T (x_k, x_{\delta,k}) \, (R + P_2 (x_k))^{-1} P_1 (x_k, x_{\delta,k}) \\
& \leq - \frac{1}{2} \|P\| \, \|f (x_k)\|^2 - \frac{1}{2} \|P\| \, \|x_{\delta,k+1}\|^2 \\
& \quad - \frac{1}{2} \|P\| \, \|x_k\|^2 - \frac{1}{2} \|P\| \, \|x_{\delta,k}\|^2
\end{aligned}
\tag{4.78}
$$

Hence, selecting P such that (4.78) is satisfied, system (2.1) with control law (4.66) guarantees asymptotic trajectory tracking along the desired trajectory $x_{\delta,k}$. It is worth noting that P and R are positive definite and symmetric matrices; thus, the existence of the inverse in (4.66) is ensured. Inverse optimality follows closely the one given in Theorem 4.1 and hence it is omitted.   ∎

4.3.1 APPLICATION TO THE BOOST CONVERTER

In this section, the inverse optimal control approach to achieve trajectory tracking, as described in the previous section, is used to control the capacitor voltage for a boost converter. Figure 4.5 shows the basic electrical circuit of the converter and the DC-to-DC converter is displayed in Figure 4.6.

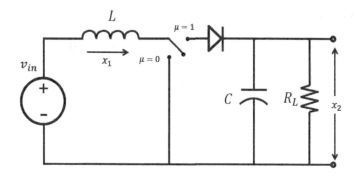

FIGURE 4.5   Boost converter circuit.

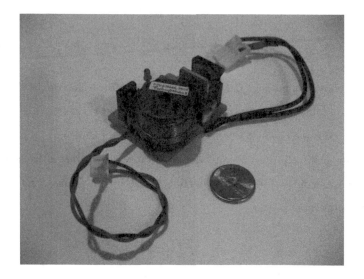

FIGURE 4.6   DC-to-DC boost converter.

4.3.1.1 Boost Converter Model

The commutated model for the boost converter can be presented as

$$
\begin{aligned}
\dot{x}_1 &= -\frac{x_2}{L}\mu + \frac{v_{in}}{L} \\
\dot{x}_2 &= -\frac{x_2}{R_L C} + \frac{x_1}{C}\mu,
\end{aligned}
\tag{4.79}
$$

where x_1 is the current across the inductor, x_2 is the voltage in the capacitor, and $\mu = \{0, 1\}$ defines the switch position; the parameters R_L, L, C, and v_{in} for the circuit are the resistance, inductance, capacitance, and source voltage, respectively, which are assumed known.

It is well known [11] that when the switching frequency is high, system (4.79) can be represented by an *average model* given as

$$
\begin{aligned}
\dot{x}_1 &= -\frac{x_2}{L}u + \frac{v_{in}}{L} \\
\dot{x}_2 &= -\frac{x_2}{R_L C} + \frac{x_1}{C}u,
\end{aligned}
\tag{4.80}
$$

where the control input $u = [0, 1]$ is the duty cycle. The value of the duty cycle determines the time in the pulse-width modulation (PWM) scheme, for which the switch is fixed at the position represented by $\mu = 1$ (see Figure 4.5) [27]. Variables x_1 and x_2 for (4.80) represent average values for current and voltage, respectively.

After discretizing by Euler approximation, the discrete-time model for the boost converter is rewritten as

$$
\begin{aligned}
x_{1,k+1} &= x_{1,k} + T\left(\frac{v_{in}}{L} - \frac{x_{2,k}}{L}u_k\right) \\
x_{2,k+1} &= x_{2,k} + T\left(-\frac{x_{2,k}}{R_L C} + \frac{x_{1,k}}{C}u_k\right),
\end{aligned}
\tag{4.81}
$$

where T is the sampling time. System (4.81) has the general affine form as (2.1) with

$$
x_k = [x_{1,k}, x_{2,k}]^T, \quad f(x_k) = \begin{bmatrix} x_{1,k} + T v_{in}/L \\ x_{2,k} - T x_{2,k}/R_L C \end{bmatrix} \quad \text{and} \quad g(x_k) = \begin{bmatrix} -T x_{2,k}/L \\ T x_{1,k}/C \end{bmatrix}, \text{ so}
$$

controller (4.66) can be used to achieve trajectory tracking.

4.3.1.2 Control Synthesis

For the case of the boost converter the aim is to control the voltage, which is done by choosing the output as

$$y = x_{2,k}. \tag{4.82}$$

However, for this output selection, the average model (4.80) becomes a nonminimum phase system. It is known that exact trajectory tracking for y as in (4.82) cannot be achieved in a nonminimum phase system [9, 10]; any control technique for accomplishing exact tracking would render the closed-loop system unstable. A basic way to solve this problem due to the nonminimum phase characteristic is to control the voltage indirectly by means of controlling the inductor current; then the output to be controlled is the inductor current variable defined as

$$y = x_{1,k}. \tag{4.83}$$

Selecting the system output to be (4.83), the system becomes minimum phase and different control techniques for trajectory tracking can be used. However, it is necessary to determine the inductor current reference to indirectly control the voltage in the capacitor. If the output voltage is constant, the stationary relation between the output voltage and the inductor current is algebraic. Such a relation can be determined from the system equilibrium point, which takes the following form:

$$\bar{x}_{1\delta} = \frac{v_{in}}{\bar{u}^2 R_L}, \qquad \bar{x}_{2\delta} = \frac{v_{in}}{\bar{u}}, \tag{4.84}$$

where $\bar{u}$, $\bar{x}_{1\delta}$ and $\bar{x}_{2\delta}$ represent steady-state values. Solving the second term in (4.84) for $\bar{u}$ and replacing $\bar{u}$ in the first one results in

$$\bar{x}_{1\delta} = \frac{\bar{x}_{2\delta}^2}{R_L v_{in}}. \tag{4.85}$$

Relation (4.85) has been extensively employed to control the boost DC-to-DC converter through the inductor current [5]. Hence, reference signals for controller (4.66), to achieve trajectory tracking for system (4.81), become

$$x_{\delta,k} = \begin{bmatrix} x_{1\delta,k} \\ x_{2\delta,k} \end{bmatrix} = \begin{bmatrix} \bar{x}_{1\delta} \\ \bar{x}_{2\delta} \end{bmatrix}.$$

4.3.1.3 Simulation Results

The parameters used for simulation are $L = 12$ mH, $R_L = 220 \ \Omega$, $C = 15 \ \mu$F, and $v_{in} = 48$ V. The sampling time is selected as $T = 33.33 \times 10^{-6}$ s. The parameters for the inverse optimal control law (4.66), in order to achieve trajectory tracking for the boost converter, are selected as $P = \begin{bmatrix} 2 & 1.5 \\ 1.5 & 2 \end{bmatrix}$ and $R = 0.5$.

Figure 4.7 presents the trajectory tracking time response by considering a constant reference for $x_{2,k} = 100$ V. The respective value for $x_{1,\delta}$ is calculated from (4.85). The initial conditions for the converter are $[x_{1,0}, x_{2,0}]^T = [0, 0]^T$. Figure 4.8 displays the applied inverse optimal control signal (4.66), which corresponds to the duty cycle.

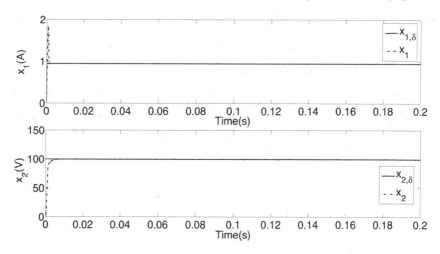

FIGURE 4.7   Trajectory tracking for a constant reference.

Figure 4.9 presents the trajectory tracking time response for a time-variant reference of x_k, with initial conditions $[x_{1,0}, x_{2,0}]^T = [0, 0]^T$. The trajectory to be tracked is $x_{2,\delta} = 100 + 30\sqrt{2} \sin(20 \pi t)$ V and the respective value for $x_{1,\delta}$ is calculated from (4.85). Figure 4.10 depicts the applied control signal (4.66), which corresponds to the

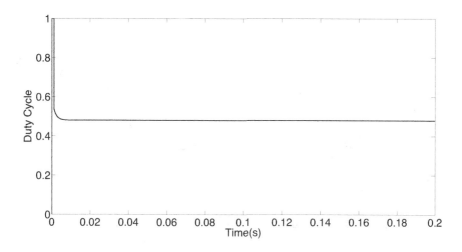

FIGURE 4.8   Control law to track a constant reference.

duty cycle to achieve trajectory tracking of a time-variant reference.

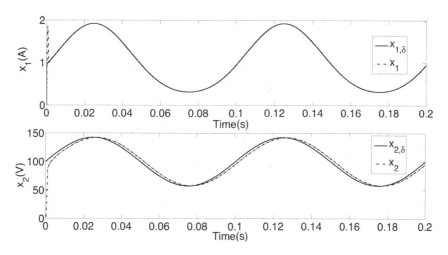

FIGURE 4.9   Trajectory tracking for a time-variant reference.

4.4   CLF-BASED INVERSE OPTIMAL CONTROL FOR A CLASS OF NONLINEAR POSITIVE SYSTEMS

Trajectory tracking for this class of nonlinear systems is presented in this section based on the CLF approach as follows.

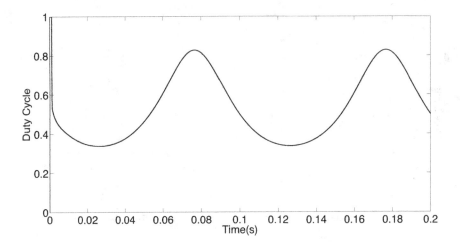

FIGURE 4.10   Control signal to track a time-variant reference.

Theorem 4.4

Consider the affine discrete-time nonlinear system (2.1). If there exists a matrix $P = P^T > 0$ such that the following inequality holds:

$$\frac{1}{2} f^T(x_k) P f(x_k) + \frac{1}{2} x_{\delta,k+1}^T P x_{\delta,k+1} - \frac{1}{2} x_k^T P x_k$$
$$- \frac{1}{2} x_{\delta,k}^T P x_{\delta,k} - \frac{1}{4} P_1^T(x_k, x_{\delta,k})(R + P_2(x_k))^{-1} P_1(x_k, x_{\delta,k})$$
$$\leq -\frac{1}{2} \|P\| \|f(x_k)\|^2 - \frac{1}{2} \|P\| \|x_{\delta,k+1}\|^2$$
$$- \frac{1}{2} \|P\| \|x_k\|^2 - \frac{1}{2} \|P\| \|x_{\delta,k}\|^2, \tag{4.86}$$

where $P_1(x_k, x_{\delta,k})$ and $P_2(x_k)$ are defined as

$$P_1(x_k, x_{\delta,k}) = \begin{cases} g^T(x_k) P\left(f(x_k) - x_{\delta,k+1}\right) & \text{for } f(x_k) \succeq x_{\delta,k+1} \\ g^T(x_k) P\left(x_{\delta,k+1} - f(x_k)\right) & \text{for } f(x_k) \preceq x_{\delta,k+1} \end{cases} \tag{4.87}$$

and

$$P_2(x_k) = \frac{1}{2} g^T(x_k) P g(x_k) \tag{4.88}$$

respectively, then system (2.1) with control law

$$u_k^* = \left| -\frac{1}{4} R^{-1} g^T (x_k) \frac{\partial z_{k+1}^T P z_{k+1}}{\partial z_{k+1}} \right|$$

$$= \left| -\frac{1}{2} (R + P_2 (x_k))^{-1} P_1 (x_k, x_{\delta,k}) \right| \qquad (4.89)$$

guarantees asymptotic trajectory tracking along the desired trajectory $x_{\delta,k}$, where $z_{k+1} = x_{k+1} - x_{\delta,k+1}$. ∎

Proof

Case 1: Consider the first case for $P_1(x_k, x_{\delta,k})$ in (4.87), that is, $P_1(x_k, x_{\delta,k}) = g^T(x_k) P (f(x_k) - x_{\delta,k+1})$. System (2.1) with control law (4.89) and (4.64), must satisfy inequality (4.63). Considering one step ahead for z_k, we have

$$\overline{V} = \frac{1}{2} z_{k+1}^T P z_{k+1} - \frac{1}{2} z_k^T P z_k + u_k^{*T} R u_k^*$$

$$= \frac{1}{2} (x_{k+1} - x_{\delta,k+1})^T P (x_{k+1} - x_{\delta,k+1}) - \frac{1}{2} (x_k - x_{\delta,k})^T P (x_k - x_{\delta,k})$$

$$+ u_k^{*T} R u_k^*. \qquad (4.90)$$

Substituting (2.1) and (4.89) in (4.90), then

$$\overline{V} = \frac{1}{2} (f(x_k) + g(x_k) u_k^* - x_{\delta,k+1})^T P (f(x_k) + g(x_k) u_k^* - x_{\delta,k+1})$$

$$- \frac{1}{2} (x - x_{\delta,k})^T P (x_k - x_{\delta,k}) + u_k^{*T} R u_k^*$$

$$= \frac{1}{2} f^T (x_k) P f(x_k) + \frac{1}{2} u_k^{*T} g^T (x_k) P g(x_k) u_k^* + \frac{1}{2} x_{\delta,k+1}^T P x_{\delta,k+1}$$

$$+ \frac{1}{2} f^T (x_k) P g(x_k) u_k^* + \frac{1}{2} u_k^{*T} g^T (x_k) P f(x_k) - \frac{1}{2} f^T (x_k) P x_{\delta,k+1}$$

$$- \frac{1}{2} x_{\delta,k+1}^T P f^T (x_k) - \frac{1}{2} u_k^T g^T (x_k) P x_{\delta,k+1} - \frac{1}{2} x_{\delta,k+1}^T P g(x_k) u_k^*$$

$$- \frac{1}{2} x_k^T P x_k - \frac{1}{2} x_{\delta,k}^T P x_{\delta,k} + \frac{1}{2} x_{\delta,k}^T P x_k + \frac{1}{2} x_k^T P x_{\delta,k} + u_k^{*T} R u_k^* \qquad (4.91)$$

By simplifying, (4.91) becomes

$$
\begin{aligned}
\overline{V} &= \frac{1}{2}f^T(x_k)\,Pf(x_k) + \frac{1}{2}u_k^{*T}g^T(x_k)\,Pg(x_k)u + \frac{1}{2}x_{\delta,k+1}^T Px_{\delta,k+1} \\
&\quad + f^T(x_k)\,Pg(x_k)u_k^* - f^T(x_k)\,Px_{\delta,k+1} - x_{\delta,k+1}^T Pg(x_k)u_k^* \\
&\quad - \frac{1}{2}x_k^T Px_k - \frac{1}{2}x_{\delta,k}^T Px_{\delta,k} + x_k^T Px_{\delta,k} + u_k^{*T}Ru_k^* \\
&= \frac{1}{2}f^T(x_k)\,Pf(x_k) + \frac{1}{2}x_{\delta,k+1}^T Px_{\delta,k+1} + x_{\delta,k+1}^T Pg(x_k)u_k \\
&\quad - x_{\delta,k+1}^T Pg(x_k)u_k^* - f^T(x_k)\,Px_{\delta,k+1} + x_k^T Px_{\delta,k} - \frac{1}{2}x_k^T Px_k \\
&\quad - \frac{1}{2}x_{\delta,k}^T Px_{\delta,k} + P_1^T(x_k, x_{\delta,k})u_k^* + u_k^{*T}P_2(x_k)u_k^* + u_k^{*T}Ru_k^* \qquad (4.92)
\end{aligned}
$$

which after using the control law (4.89), (4.92) results in

$$
\begin{aligned}
\overline{V} &= \frac{1}{2}f^T(x_k)\,Pf(x_k) + \frac{1}{2}x_{\delta,k+1}^T Px_{\delta,k+1} - \frac{1}{2}x_k^T Px_k - \frac{1}{2}x_{\delta,k}^T Px_{\delta,k} \\
&\quad - \frac{1}{4}P_1^T(x_k, x_{\delta,k})\,(R+P_2(x_k))^{-1}P_1(x_k, x_{\delta,k}) \\
&\quad - f^T(x_k)\,Px_{\delta,k+1} + x_k^T Px_{\delta,k} \qquad (4.93)
\end{aligned}
$$

Analyzing the sixth and seventh RHS terms of (4.93) by using the inequality $X^TY + Y^TX \le X^T\Lambda X + Y^T\Lambda^{-1}Y$ proved in [30], which is valid for any vector $X \in \mathbb{R}^{n\times 1}$, then for the sixth RHS term of (4.93), we have

$$
\begin{aligned}
f^T(x_k)Px_{\delta,k+1} &\le \frac{1}{2}\left[f^T(x_k)\,Pf(x_k) + \left(Px_{\delta,k+1}\right)^T P^{-1}\left(Px_{\delta,k+1}\right)\right] \\
&\le \frac{1}{2}\left[f^T(x_k)\,Pf(x_k) + x_{\delta,k+1}^T Px_{\delta,k+1}\right] \\
&\le \frac{1}{2}\|P\|\,\|f\|^2 + \frac{1}{2}\|P\|\,\|x_{\delta,k+1}\|^2. \qquad (4.94)
\end{aligned}
$$

The seventh RHS term of (4.93) becomes

$$
\begin{aligned}
x_k^T Px_{\delta,k} &\le \frac{1}{2}\left[x_k^T Px_k + \left(Px_{\delta,k}\right)^T P^{-1}\left(Px_{\delta,k}\right)\right] \\
&\le \frac{1}{2}\left[x_k^T Px_k + \left(Px_{\delta,k}\right)^T P\left(x_{\delta,k}\right)\right]. \qquad (4.95)
\end{aligned}
$$

From (4.95), the following expression holds.

$$\frac{1}{2}\left[x_k^T P x_k + \left(P x_{\delta,k}\right)^T P^{-1}\left(P x_{\delta,k}\right)\right] \leq \frac{1}{2}\|P\|\,\|x_k\|^2 + \frac{1}{2}\|P\|\,\|x_{\delta,k}\|. \tag{4.96}$$

Substituting (4.94) and (4.96) into (4.93), then

$$\begin{aligned}
\overline{V} = \ &\frac{1}{2}f^T\left(x_k\right) P f\left(x_k\right) + \frac{1}{2}x_{\delta,k+1}^T P x_{\delta,k+1} - \frac{1}{2}x_k^T P x_k \\
&-\frac{1}{2}x_{\delta,k}^T P x_{\delta,k} - \frac{1}{4}P_1^T\left(x_k, x_{\delta,k}\right)\left(R + P_2\left(x_k\right)\right)^{-1} P_1\left(x_k, x_{\delta,k}\right) \\
&+\frac{1}{2}\|P\|\,\|f\left(x_k\right)\|^2 + \frac{1}{2}\|P\|\,\|x_{\delta,k+1}\|^2 \\
&+\frac{1}{2}\|P\|\,\|x_k\|^2 + \frac{1}{2}\|P\|\,\|x_{\delta,k}\|^2
\end{aligned} \tag{4.97}$$

In order to achieve asymptotic stability, it is required that $\overline{V} \leq 0$, then from (4.97) the next inequality is formulated

$$\begin{aligned}
\frac{1}{2}f^T\left(x_k\right) & P f\left(x_k\right) + \frac{1}{2}x_{\delta,k+1}^T P x_{\delta,k+1} - \frac{1}{2}x_k^T P x_k \\
&-\frac{1}{2}x_{\delta,k}^T P x_{\delta,k} - \frac{1}{4}P_1^T\left(x_k, x_{\delta,k}\right)\left(R + P_2\left(x_k\right)\right)^{-1} P_1\left(x_k, x_{\delta,k}\right) \\
&\leq -\frac{1}{2}\|P\|\,\|f\left(x_k\right)\|^2 - \frac{1}{2}\|P\|\,\|x_{\delta,k+1}\|^2 \\
&\quad -\frac{1}{2}\|P\|\,\|x_k\|^2 - \frac{1}{2}\|P\|\,\|x_{\delta,k}\|^2.
\end{aligned} \tag{4.98}$$

Hence, selecting P such that (4.98) is satisfied, system (2.1) with control law (4.89) guarantees asymptotic trajectory tracking along the desired trajectory $x_{\delta,k}$. It is worth noting that P and R are positive definite and symmetric matrices; thus, the existence of the inverse in (4.89) is ensured.

Case 2: $P_1\left(x_k, x_{\delta,k}\right) = g^T\left(x_k\right) P\left(x_{\delta,k+1} - f\left(x_k\right)\right)$. It can be derived in the same way as in Case 1. Finally the proposed inverse optimal control law is given as

$$u_k^* = \left|-\frac{1}{2}\left(R + P_2\left(x_k\right)\right)^{-1} P_1\left(x_k, x_{\delta,k}\right)\right|, \tag{4.99}$$

which ensures that $P_1\left(x_k, x_{\delta,k}\right)$ satisfies (4.87).

Inverse optimality follows closely the one given in Theorem 4.1 and hence it is omitted.   ∎

4.5 SPEED-GRADIENT FOR THE INVERSE OPTIMAL CONTROL

In Section 4.1, a CLF candidate such as $V(x_k) = \frac{1}{2}x_k^T P x_k$ is proposed in order to solve the inverse optimal control as established in Definition 4.1, for which an adequate selection of the fixed matrix P must be done such that condition (4.7) is fulfilled. In this section, we propose using the speed-gradient algorithm to calculate this matrix P in a recursive way to ensure the fulfillment of condition (4.7). Then, a CLF candidate $V(x_k)$ described by

$$V(x_k) = \frac{1}{2}x_k^T P_k x_k, \qquad P_k = P_k^T > 0 \qquad\qquad (4.100)$$

is proposed for control law (4.1) in order to guarantee stability for the equilibrium point $x_k = 0$ of system (2.1). Stability will be achieved by defining an appropriate matrix P_k. Moreover, it will be established that control law (4.1) based on (4.100) optimizes the cost functional (2.2).

Consequently, by considering $V(x_k) = V_c(x_k)$ as in (4.100), the control law (4.1) takes the following form:

$$u_k^* = -\frac{1}{2}\left(R + \frac{1}{2}g^T(x_k) P_k g(x_k) \right)^{-1} g^T(x_k) P_k f(x_k). \qquad\qquad (4.101)$$

It is worth pointing out that P_k and R are positive definite matrices; thus, the existence of the inverse in (4.101) is assured. To determine P_k, which ensures stability of the equilibrium point $x_k = 0$ of system (2.1) with (4.101), in this section we propose using the speed-gradient (SG) algorithm.

4.5.1 SPEED-GRADIENT ALGORITHM

The goal of the discrete-time SG algorithm, proposed in [6], is to determine a parameter p, which ensures the following goal:

$$\mathcal{Q}(p) \leq \Delta, \qquad \text{for } k \geq k^*, \tag{4.102}$$

where $\mathcal{Q}$ is a positive definite goal function, Δ is a positive constant considered as a threshold, and $k^* \in \mathbb{Z}^+$ is the time step at which the goal is achieved. Related results on the continuous-time and discrete-time speed-gradient algorithms and their applications are given in [6, 24] and references therein.

Control law (4.101) at every time step depends on the matrix P_k, which is defined as

$$P_k = p_k P',$$

where $P' = P'^T > 0$ is a given constant matrix and p_k is a scalar parameter to be determined by the SG algorithm. Then, (4.101) is transformed into

$$u_k^* = -\frac{p_k}{2} \left(R + \frac{p_k}{2} g^T(x_k) P' g(x_k) \right)^{-1} g^T(x_k) P' f(x_k). \tag{4.103}$$

The SG algorithm is now reformulated for inverse optimal control.

Definition 4.4: SG Goal Function Consider a time-varying parameter $p_k \in \mathbb{R}^+$. The positive definite $\mathscr{C}^1$ function $\mathcal{Q} : \mathbb{R}^n \times \mathbb{R}^+ \rightarrow \mathbb{R}^+$ given as

$$\mathcal{Q}(x_k, p_k) = V_{sg}(x_{k+1}), \tag{4.104}$$

where $V_{sg}(x_{k+1}) = \frac{1}{2} x_{k+1}^T P' x_{k+1}$, with $x_{k+1} = f(x_k) + g(x_k) u_k^*$, is called the SG goal function for system (2.1) with control law (4.103).

The SG goal function is defined as in (4.104) in such a way that the convexity property of $\mathcal{Q}(x_k, p_k)$ for p_k is guaranteed; then there exist an optimal value p^* for p_k and a positive constant ε^* such that $\mathcal{Q}(x_k, p^*) \leq \varepsilon^*$ [6]. In Theorem 4.5 below, this SG goal function is used to construct a Lyapunov function for the closed-loop system.

Definition 4.5: SG Control Goal The SG control goal for system (2.1) with (4.103) is defined as

$$\mathcal{Q}(x_k, p_k) \leq \Delta(x_k), \qquad \text{for} \quad k \geq k^*, \tag{4.105}$$

where p_k is the value such that (4.105) is fulfilled, with

$$\Delta(x_k) = V_{sg}(x_k) - \frac{1}{p_k} u_k^{*T} R u_k^*, \tag{4.106}$$

$V_{sg}(x_k) = \frac{1}{2} x_k^T P' x_k$ and $k^* \in \mathbb{Z}_+$ is the time step at which the SG control goal is achieved.

Comment 4.5 Solution p_k must guarantee that $V_{sg}(x_k) > \frac{1}{p_k} u_k^{*T} R u_k^*$ in order to obtain a positive definite function $\Delta(x_k)$.

Let us state the first contribution of the section as follows.

Lemma 4.1

Consider the discrete-time nonlinear system (2.1) with (4.103) as input. Let $\mathcal{Q}$ be an SG goal function as defined in (4.104). Let $\bar{p}$ be a positive constant, $\Delta(x_k)$ be a positive definite function with $\Delta(0) = 0$, and assume that there exist positive constants

p^* and ε^* such that the following control goal is achievable [6]:

$$\mathcal{Q}(x_k, p^*) \leq \varepsilon^* \ll \Delta(x_k). \tag{4.107}$$

Then, for any initial condition $p_0 > 0$, there exists a $k^* \in \mathbb{Z}^+$ such that the SG control goal (4.105) is achieved by means of the following dynamic variation of parameter p_k:

$$p_{k+1} = p_k - \gamma_{d,k} \nabla_p \mathcal{Q}(x_k, p_k), \tag{4.108}$$

with

$$\gamma_{d,k} = \gamma_c \, \delta_k \, |\nabla_p \mathcal{Q}(x_k, p_k)|^{-2}, \qquad 0 < \gamma_c \leq 2\Delta(x_k)$$

and

$$\delta_k = \begin{cases} 1 & for \quad \mathcal{Q}(x_k, p_k) > \Delta(x_k) \\ 0 & otherwise. \end{cases} \tag{4.109}$$

Finally, for $k \geq k^*$, p_k becomes a constant denoted by $\bar{p}$ and the SG algorithm terminates. ∎

Proof

Along the lines of [6], the proof is based on the case for which $\mathcal{Q}(x_k, p_k) > \Delta(x_k)$, and therefore $\delta_k = 1$. Let us consider the positive definite Lyapunov function $V_p(p_k) = |p_k - p^*|^2$. Then, the respective Lyapunov difference is given as

$$\begin{aligned} \Delta V_p(p_k) &= |p_{k+1} - p^*|^2 - |p_k - p^*|^2 \\ &= (p_{k+1} - p_k)^T [(p_{k+1} - p_k) + 2(p_k - p^*)] \\ &= -\gamma_{d,k} \nabla_p \mathcal{Q}(x_k, p_k) [-\gamma_{d,k} \nabla_p \mathcal{Q}(x_k, p_k) + 2(p_k - p^*)] \tag{4.110} \end{aligned}$$

Due to convexity of the SG goal function (4.104) for p_k,

$$(p^* - p_k)^T \nabla_p \mathcal{Q}(x_k, p_k) \leq \varepsilon^* - \Delta(x_k) < 0, \tag{4.111}$$

where $\nabla_p \mathscr{Q}(x_k, p_k)$ denotes the gradient of $\mathscr{Q}(x_k, p_k)$ with respect to p_k. Based on (4.111), (4.110) becomes

$$
\begin{aligned}
\Delta V_p(p_k) & \leq -2\gamma_{d,k} \left(\Delta(x_k) - \varepsilon^* \right) + \gamma_{d,k}^2 \left| \nabla_p \mathscr{Q}(x_k, p_k) \right|^2 \\
& \leq -2\gamma_c \delta_k \left(\Delta(x_k) - \varepsilon^* \right) \left| \nabla_p \mathscr{Q}(x_k, p_k) \right|^{-2} \\
& \quad + \gamma_c^2 \delta_k^2 \left| \nabla_p \mathscr{Q}(x_k, p_k) \right|^{-4} \left| \nabla_p \mathscr{Q}(x_k, p_k) \right|^2 \\
& = -\frac{\gamma_c \left[2\Delta(x_k) \left(1 - \left(\varepsilon^*/\Delta(x_k) \right) \right) - \gamma_c \right]}{\left| \nabla_p \mathscr{Q}(x_k, p_k) \right|^2}.
\end{aligned}
$$

From (4.107), $1 - \left(\varepsilon^*/\Delta(x_k) \right) \approx 1$, hence

$$
\begin{aligned}
\Delta V_p(p_k) & \approx -\frac{\gamma_c \left[2\Delta(x_k) - \gamma_c \right]}{\left| \nabla_p \mathscr{Q}(x_k, p_k) \right|^2} \\
& < 0.
\end{aligned}
$$

Thus, boundedness of p_k is guaranteed if $0 < \gamma_c \leq 2\Delta(x_k)$. Finally, when $k \geq k^*$, then $\delta_k = 0$, which means the algorithm terminates; at this point $\mathscr{Q}(x_k, p_k) \leq \Delta(x_k)$, then p_k becomes a constant value denoted by $\bar{p}$ (i.e., $p_k = \bar{p}$). ∎

Note that the gradient $\nabla_p \mathscr{Q}(x_k, p_k)$ in (4.108) is reduced to being the partial derivative of $\mathscr{Q}(x_k, p_k)$ with respect to p_k, as $\frac{\partial}{\partial p_k} \mathscr{Q}(x_k, p_k)$.

Comment 4.6 Parameter γ_c in (4.108) is selected such that the solution p_k ensures the requirement $V_{sg}(x_k) > \frac{1}{p_k} u_k^T R u_k$ in Comment 4.5. Then, we have a positive definite function $\Delta(x_k)$.

Comment 4.7 With $\mathscr{Q}(x_k, p_k)$ as defined in (4.104), the dynamic variation of parameter p_k in (4.108) results in

$$p_{k+1} = p_k + 8\gamma_{d,k}\frac{f^T(x_k)P'g(x_k)R^2g^T(x_k)f(x_k)}{\left(2R + p_kg^T(x_k)P'g(x_k)\right)^3}$$

which is positive for all time steps k if $p_0 > 0$. Therefore positiveness for p_k is ensured and the requirement $P_k = P_k^T > 0$ for (4.100) is guaranteed.

When the SG control goal (4.105) is achieved, then $p_k = \bar{p}$ for $k \geq k^*$. Thus, matrix P_k in (4.101) is considered constant and $P_k = P$, where P is computed as $P = \bar{p}P'$, with P' a design positive definite matrix. Under these constraints, we obtain

$$\begin{aligned}\alpha(x_k) \quad &:= \quad u_k^* \\ &= \quad -\frac{1}{2}\left(R + \frac{1}{2}g^T(x_k)Pg(x_k)\right)^{-1}g^T(x_k)Pf(x_k). \quad (4.112)\end{aligned}$$

Figure 4.11 presents the flow diagram for the proposed SG algorithm.

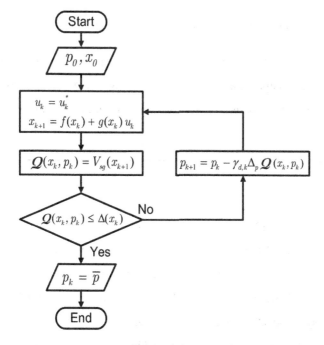

FIGURE 4.11   Speed-gradient algorithm flow diagram.

4.5.2 SUMMARY OF THE PROPOSED SG ALGORITHM TO CALCULATE PA-
RAMETER P_K

Considering the closed-loop system (2.1) with (4.103) as input, we obtain

$$x_{k+1} = f(x_k) - \frac{p_k}{2} g(x_k) \left(R + \frac{p_k}{2} g^T(x_k) P' g(x_k) \right)^{-1} g^T(x_k) P' f(x_k).$$

Then, we propose the SG goal function

$$\mathcal{D}_k(p_k) = x_{k+1}^T x_{k+1}.$$

The dynamic variation of parameter p_k is established as

$$p_{k+1} = p_k - \gamma \nabla_p \mathcal{D}_k(p_k), \qquad p_0 = p(0).$$

Finally, when condition (4.105) is fulfilled, the SG algorithm finishes.

4.5.3 SG INVERSE OPTIMAL CONTROL

Once the control law (4.112) has been determined, we proceed to demonstrate that it ensures stability and optimality for (2.1) without solving the HJB equation (2.10).

Thus, the second contribution of this section is stated as the following theorem.

Theorem 4.5

Consider that system (2.1) with (4.103) has achieved the SG control goal (4.105) by means of (4.108). Let $V(x_k) = \frac{1}{2} x_k^T P x_k$ be a Lyapunov function candidate with $P = P^T > 0$. Then, control law (4.112) is an inverse optimal control law, in accordance with Definition 4.1, which ensures that the equilibrium point $x_k = 0$ of system (2.1) is globally asymptotically stable. Moreover, with $V(x_k) = \frac{1}{2} x_k^T P x_k$ as a CLF and $P = \bar{p} P'$, control law (4.112) is inverse optimal in the sense that it minimizes the cost functional given by

$$\mathcal{J} = \sum_{k=0}^{\infty} \left(l(x_k) + u_k^T R \, u_k \right), \tag{4.113}$$

where

$$l(x_k) := -\overline{V} \tag{4.114}$$

with $\overline{V}$ defined as

$$\overline{V} = V(x_{k+1}) - V(x_k) + \alpha^T(x_k) R \alpha(x_k).$$

∎

Proof

Considering that system (2.1), with control law (4.103) and the SG algorithm (4.108) achieves the SG control goal (4.105) for $k \geq k^*$ (Lemma 4.1), then (4.105) can be rewritten as

$$\begin{aligned}
V_{sg}(x_{k+1}) - V_{sg}(x_k) &+ \frac{1}{\bar{p}} \alpha^T(x_k) R \alpha(x_k) \\
&= \frac{1}{2} x_{k+1}^T P' x_{k+1} - \frac{1}{2} x_k^T P' x_k + \frac{1}{\bar{p}} \alpha^T(x_k) R \alpha(x_k) \\
&\leq 0. \tag{4.115}
\end{aligned}$$

Multiplying (4.115) by the positive constant $\bar{p}$ of Lemma 4.1, and using the SG goal function as a Lyapunov function given as $V(x_k) = \bar{p} V_{sg}(x_k)$ for the closed-loop system, we obtain

$$\begin{aligned}
\overline{V} &:= \frac{\bar{p}}{2} x_{k+1}^T P' x_{k+1} - \frac{\bar{p}}{2} x_k^T P' x_k + \alpha^T(x_k) R \alpha(x_k) \\
&= \frac{1}{2} x_{k+1}^T P x_{k+1} - \frac{1}{2} x_k^T P x_k + \alpha^T(x_k) R \alpha(x_k) \\
&= V(x_{k+1}) - V(x_k) + \alpha^T(x_k) R \alpha(x_k) \\
&\leq 0. \tag{4.116}
\end{aligned}$$

From (4.116), obviously $V(x_{k+1}) - V(x_k) < 0$ for all $x_k \neq 0$ with $V(x_k)$ a positive definite and radially unbounded function, then global asymptotic stability is achieved in accordance with Theorem 2.1.

When function $-l(x_k)$ is set to be the RHS of (4.116), then

$$
\begin{aligned}
l(x_k) \quad &:= \quad -\overline{V} \\
&= \quad -(V(x_{k+1}) - V(x_k)) - \alpha^T(x_k)R\alpha(x_k) \\
&\geq \quad 0, \qquad \forall x_k \neq 0.
\end{aligned}
\tag{4.117}
$$

Consequently, $V(x_k) = \frac{1}{2}x_k^T P x_k$ as a CLF is a solution of the HJB equation (2.10) for $k \geq k^*$.

In order to obtain the optimal value function for the cost functional (4.113), we proceed as in Theorem 4.1. ∎

As established in [31], to use a CLF for the inverse optimal control approach, the entries of P' in $V(x_k)$ are selected such that $\Delta V(x_k, u_k)$ is negative definite, which considers the flexibility provided by the control term $g(x_k)u_k$. Additionally, once the condition $\Delta V(x_k, u_k) < 0$ is fulfilled, adequate values for the entries of P' are a matter of trial and error such that good performance is achieved for the system dynamics. These arguments motivated us to explore the use of the SG algorithm for the selection of matrix P'.

It is worth mentioning that the CLF approach for control synthesis has been applied successfully to systems for which a CLF can be established, such as feedback linearizable, strict feedback, and feed-forward ones [8, 28]. However, systematic techniques for determining CLFs do not exist for general nonlinear systems [28].

In this work, we refer to (2.2) as a cost functional due to the fact that a weighting for both the state and the control input can be established. The weighting of the control input in (2.2) is selected directly by R, while for the state it is analyzed as follows:

$$
\begin{aligned}
l(x_k) &= -\left[V(x_{k+1}) - V(x_k) + \alpha^T(x_k) R \alpha(x_k)\right] \\
&= -\frac{f^T(x_k) P f(x_k) + 2 f^T(x_k) P g(x_k) \alpha(x_k)}{2} \\
&\quad + \frac{x_k^T P x_k - \alpha^T(x_k) g^T(x_k) P g(x_k) \alpha(x_k)}{2} - \alpha^T(x_k) R \alpha(x_k) \\
&= \frac{1}{2}\left(x_k^T P x_k - f^T(x_k) P f(x_k)\right) + \frac{1}{2} f^T(x_k) P g(x_k) \\
&\quad \times \left(R + \frac{1}{2} g^T(x_k) P g(x_k)\right)^{-1} g^T(x_k) P f(x_k) - \frac{1}{4} f^T(x_k) P g(x_k) \\
&\quad \times \left(R + \frac{1}{2} g^T(x_k) P g(x_k)\right)^{-1} g^T(x_k) P f(x_k) \\
&= \frac{1}{2}\left(x_k^T P x_k - f^T(x_k) P f(x_k)\right) + \frac{1}{4} f^T(x_k) P g(x_k) \\
&\quad \times \left(R + \frac{1}{2} g^T(x_k) P g(x_k)\right)^{-1} g^T(x_k) P f(x_k)
\end{aligned}
$$

which can be rewritten as

$$
\begin{aligned}
l(x_k) &= \frac{1}{2} f^T(x_k) \left[\frac{1}{2} P g(x_k)\left(R + \frac{1}{2} g^T(x_k) P g(x_k)\right)^{-1} g^T(x_k) P - P\right] f(x_k) \\
&\quad + \frac{1}{2} x_k^T P x_k \\
&= \frac{1}{2}\begin{bmatrix} f(x_k) & x_k \end{bmatrix}^T \\
&\quad \times \begin{bmatrix} \frac{1}{2} P g(x_k)\left(R + \frac{1}{2} g^T(x_k) P g(x_k)\right)^{-1} g^T(x_k) P - P & 0 \\[2mm] 0 & P \end{bmatrix} \\
&\quad \times \begin{bmatrix} f(x_k) \\ x_k \end{bmatrix}.
\end{aligned}
$$

Hence, selecting an adequate P for $l(x_k)$, a weight term for x_k can be obtained.

4.5.3.1 Example

We synthesize an inverse optimal control law for a discrete-time second order non-linear system (unstable for $u_k = 0$) of the form (2.1) with

$$f(x_k) = \begin{bmatrix} x_{1,k}^2 x_{2,k} - 0.8 x_{2,k} \\ x_{1,k}^2 + 1.8 x_{2,k} \end{bmatrix} \tag{4.118}$$

and

$$g(x_k) = \begin{bmatrix} 0 \\ -2 + cos(x_{2,k}) \end{bmatrix}. \tag{4.119}$$

According to (4.112), the inverse optimal control law is formulated as

$$u_k^* = -\frac{1}{2} \left(R + \frac{1}{2} g^T(x_k) P_k g(x_k) \right)^{-1} g^T(x_k) P_k f(x_k),$$

where the positive definite matrix $P_k = p_k P'$ is calculated by the SG algorithm, with P' as the identity matrix, that is,

$$\begin{aligned} P_k &= p_k P' \\ &= p_k \begin{bmatrix} 1 & 0 \\ 0 & 1 \end{bmatrix} \end{aligned}$$

and R is a constant term selected as

$$R = 0.2.$$

For this simulation, the selection of the identity matrix for P' is sufficient to ensure asymptotic stability. The state penalty term $l(x_k)$ in (4.113) is calculated according to (4.114). The phase portrait for this unstable open-loop ($u_k = 0$) system with initial conditions $\chi_0 = [2, -2]^T$ is displayed in Figure 4.12. Figure 4.13 shows the time evolution of x_k for this system with initial conditions $x_0 = [2, -2]^T$ under the action of the proposed control law. This figure also includes the applied inverse optimal control

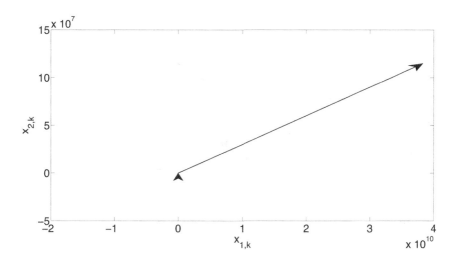

FIGURE 4.12   Open-loop unstable phase portrait.

law, which achieves asymptotic stability; the respective phase portrait is portrayed in Figure 4.14. Figure 4.15 displays the SG algorithm solution p_k; the evaluation of the cost functional $\mathscr{J}$ is also shown in this figure.

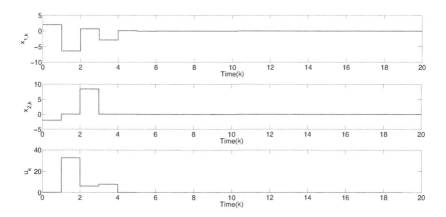

FIGURE 4.13   Stabilization of a nonlinear system using the speed-gradient algorithm.

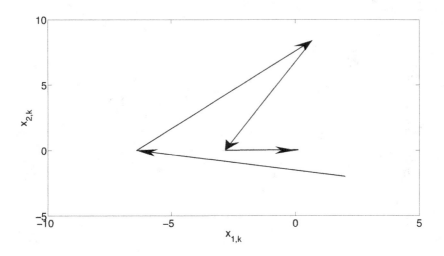

FIGURE 4.14   Phase portrait for the stabilized system using the speed-gradient algorithm.

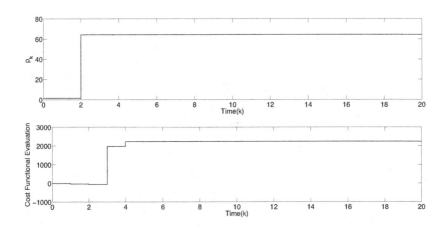

FIGURE 4.15 p_k and $\mathscr{J}$ time evolution.

4.5.4 APPLICATION TO THE INVERTED PENDULUM ON A CART

The proposed inverse optimal control is illustrated by stabilizing the inverted pendulum on a cart at the upright position [22] (see Figure 4.17), which is difficult to control due to the fact that it is an underactuated system with $\overrightarrow{F}$ the only control input. The control scheme for the pendulum on a cart could be used in applications such as the Segway personal transporter (see Figure 4.16). The dynamics of the inverted

FIGURE 4.16 Segway personal transporter.

pendulum are given as [22]

$$
\begin{aligned}
\dot{x} &= v_x \\
\dot{v}_x &= \frac{ml\omega^2 \sin\theta - mg\sin\theta\cos\theta + \overrightarrow{F}}{M + m\sin^2\theta} \\
\dot{\theta} &= \omega \\
\dot{\omega} &= \frac{-ml\omega^2 \sin\theta\cos\theta + (M+m)g\sin\theta - \overrightarrow{F}\cos\theta}{Ml + ml\sin^2\theta},
\end{aligned}
\tag{4.120}
$$

where x is the car position, v_x is the car velocity, θ is the pendulum angle, ω is the angular velocity, M is the mass of the car, m is the point mass attached at the end of the pendulum, l is the length of the pendulum, g is the gravity constant, and $\overrightarrow{F}$ is the force applied to the cart.

After discretizing by Euler approximation,[1] the discrete-time model for the inverted pendulum on a cart is rewritten as

$$
\begin{aligned}
x_{k+1} &= x_k + T v_{x,k} \\
v_{x,k+1} &= v_{x,k} + T \left(\frac{m l \omega_k^2 \sin \theta_k - m g \sin \theta_k \cos \theta_k}{M + m \sin^2 \theta_k} \right) + \frac{T}{M + m \sin^2 \theta_k} \overrightarrow{F}_k \\
\theta_{k+1} &= \theta_k + T \omega_k \\
\omega_{k+1} &= \omega_k + T \left(\frac{-m l \omega_k^2 \sin \theta_k \cos \theta_k + (M+m) g \sin \theta_k}{M l + m l \sin^2 \theta_k} \right) \\
&\quad + \frac{-T \cos \theta_k}{M l + m l \sin^2 \theta_k} \overrightarrow{F}_k,
\end{aligned}
\tag{4.121}
$$

where T is the sampling time.

System (4.121) can be presented in a general affine form as

$$
x_{k+1} = f(x_k) + g(x_k) \overrightarrow{F}_k, \tag{4.122}
$$

where $x_k = \left[x_k, v_{x,k}, \theta_k, \omega_k \right]^T$, and the inverse optimal control law (4.112) is applied for this system as $\overrightarrow{F}_k = \alpha(x_k)$.

4.5.4.1 Simulation Results

The parameters used for simulation are $M = 3$ kg, $m = 1$ kg, $l = 0.5$ m, $g = 9.81$ m/s^2, and the sampling time is $T = 0.001$ s. The initial conditions are $\left[x_0, v_{x,0}, \theta_0, \omega_0 \right]^T =$

---

[1]For the ordinary differential equation $\frac{dz}{dt} = f(z)$, the Euler discretization is defined as $\frac{z_{k+1} - z_k}{T} = f(z_k)$, such that $z_{k+1} = z_k + T f(z_k)$, where T is the sampling time [16, 20].

$[0,0,0.5,0]^T$. For this simulation, the selection of the matrix P' in (4.112) is done as

$$
P' = \begin{bmatrix}
2.81 & 1.80 & 1.70 & 1.20 \\
1.80 & 2.81 & 2.80 & 2.79 \\
1.70 & 2.80 & 3.00 & 2.90 \\
1.20 & 2.79 & 2.90 & 3.00
\end{bmatrix}
$$

which is sufficient to ensure asymptotic stability on the desired angle position. The initial condition for p_k in the SG algorithm is $p_0 = 5$. Matrix R in (4.113) is selected as

$$
R = 0.5.
$$

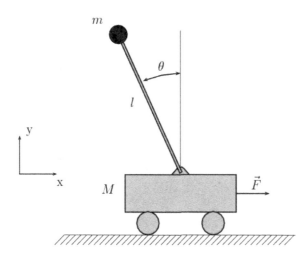

FIGURE 4.17   Inverted pendulum on a cart.

Figure 4.18 presents the time evolution of x_k, $v_{x,k}$, θ_k, and ω_k. There, it can be seen that the stabilization of the inverted pendulum in the upright position ($\theta = 0$ rad) is achieved. Figure 4.19 displays the applied inverse optimal control law and the evaluation of the respective cost functional.

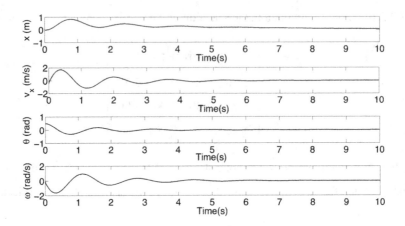

FIGURE 4.18   Stabilized inverted pendulum time response.

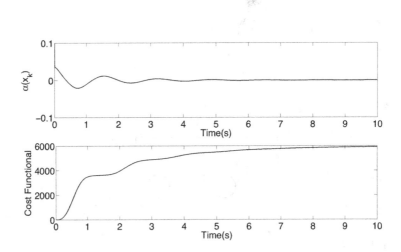

FIGURE 4.19   Control signal and cost functional time evolution for the inverted pendulum.

4.6 SPEED-GRADIENT ALGORITHM FOR TRAJECTORY TRACKING

In this section, instead of determining a fixed matrix P as proposed in (4.64), a time-varying matrix P_k is calculated by using the speed-gradient algorithm, which ensures trajectory tracking of x_k for system (2.1) along the desired trajectory $x_{\delta,k}$.

As in Section 4.5, the control goal function is established as

$$\mathscr{Q}(z_{k+1}) \leq \Delta, \qquad \text{for } k \geq k^*, \tag{4.123}$$

where $\mathscr{Q}$ is a control goal function, a constant $\Delta > 0$, and $k^* \in \mathbb{Z}^+$ is the time at which the control goal is achieved.

Let us define the matrix P_k at every time k as

$$P_k = p_k \overline{P},$$

where p_k is a scalar parameter to be adjusted by the SG algorithm, $\overline{P} = K^T P' K$ with K an additional diagonal gain matrix of appropriate dimension introduced to modify the convergence rate of the tracking error, and $P' = P'^T > 0$, a design constant matrix of appropriate dimension. Then, in a similar way as established in Section 4.3, the inverse optimal control law, which uses the SG algorithm, becomes

$$u_k = -\frac{p_k}{2}\left(R + \frac{p_k}{2}g^T(x_k)\overline{P}g(x_k)\right)^{-1}g^T(x_k)\overline{P}\left(f(x_k) - x_{\delta,k+1}\right). \tag{4.124}$$

The SG algorithm is now reformulated for trajectory tracking inverse optimal control.

Definition 4.6: SG Goal Function for Trajectory Tracking Consider a time-varying parameter $p_k \in \mathscr{P} \subset \mathbb{R}^+$, with $p_k > 0$ for all k, and $\mathscr{P}$ is the set of admissible values for p_k. A nonnegative $\mathscr{C}^1$ function $\mathscr{Q}: \mathbb{R}^n \times \mathbb{R} \to \mathbb{R}$ of the form

$$\mathscr{Q}(z_k, p_k) = V_{sg}(z_{k+1}), \tag{4.125}$$

where $V_{sg}(z_{k+1}) = \frac{1}{2} z_{k+1}^T P' z_{k+1}$, is referred to as the SG goal function for system (2.1), with $z_{k+1} = x_{k+1} - x_{\delta,k+1}$, x_{k+1} as defined in (2.1), control law (4.124), and desired reference $x_{\delta,k+1}$. We define $\mathscr{Q}_k(p) := \mathscr{Q}(z_k, p_k)$.

Definition 4.7: SG Control Goal for Trajectory Tracking Consider a constant $p^* \in \mathscr{P}$. The SG control goal for system (2.1) with (4.124) is defined as finding p_k so that the SG goal function $\mathscr{Q}_k(p)$ as defined in (4.125) fulfills

$$\mathscr{Q}_k(p) \leq \Delta(z_k), \qquad \text{for} \quad k \geq k^*, \tag{4.126}$$

where

$$\Delta(z_k) = V_{sg}(z_k) - \frac{1}{p_k} u_k^T R u_k \tag{4.127}$$

with $V_{sg}(z_k) = \frac{1}{2} z_k^T P' z_k$ and u_k as defined in (4.124); $k^* \in \mathbb{Z}^+$ is the time at which the SG control goal is achieved.

Comment 4.8 Solution p_k must guarantee that $V_{sg}(z_k) > \frac{1}{p_k} u_k^T R u_k$ in order to obtain a positive definite function $\Delta(z_k)$.

The SG algorithm is used to compute p_k in order to achieve the SG control goal defined above.

Lemma 4.2

Consider a discrete-time nonlinear system of the form (2.1) with (4.124) as input. Let $\mathscr{Q}$ be an SG goal function as defined in (4.125), and denoted by $\mathscr{Q}_k(p)$. Let $\bar{p}, p^* \in \mathscr{P}$ be positive constant values, $\Delta(z_k)$ be a positive definite function with $\Delta(0) = 0$, and ε^* be a sufficiently small positive constant. Assume the following:

- A1. There exist p^* and ε^* such that

$$\mathcal{Q}_k(p^*) \leq \varepsilon^* \ll \Delta(z_k) \quad \text{and} \quad 1 - \varepsilon^*/\Delta(z_k) \approx 1. \tag{4.128}$$

- A2. For all $p_k \in \mathcal{P}$:

$$(p^* - p_k)^T \nabla_p \mathcal{Q}_k(p) \leq \varepsilon^* - \Delta(z_k) < 0, \tag{4.129}$$

where $\nabla_p \mathcal{Q}_k(p)$ denotes the gradient of $\mathcal{Q}_k(p)$ with respect to p_k.

Then, for any initial condition $p_0 > 0$, there exists a $k^* \in \mathbb{Z}^+$ such that the SG control goal (4.126) is achieved by means of the following dynamical variation of parameter p_k:

$$p_{k+1} = p_k - \gamma_{d,k} \nabla_p \mathcal{Q}_k(p), \tag{4.130}$$

with

$$\gamma_{d,k} = \gamma_c \, \delta_k \left| \nabla_p \mathcal{Q}_k(p) \right|^{-2}, \qquad 0 < \gamma_c \leq 2\Delta(z_k)$$

and

$$\delta_k = \begin{cases} 1 & \text{for} \quad Q(p_k) > \Delta(z_k) \\ 0 & \text{otherwise.} \end{cases} \tag{4.131}$$

Finally, for $k \geq k^*$, p_k becomes a constant value denoted by $\bar{p}$ and the SG algorithm is completed. ■

Proof

It follows closely the proof of Lemma 4.1.   ■

Comment 4.9 Parameter γ_c in (4.130) is selected such that solution p_k ensures the requirement $V_{sg}(z_k) > \frac{1}{p_k} u_k^T R u_k$ in Comment 4.8. Then, we have a positive definite function $\Delta(z_k)$.

When the SG control goal (4.126) is achieved, then $p_k = \bar{p}$ for $k \geq k^*$. Thus, matrix P_k is considered constant, that is, $P_k = P$, where P is computed as $P = \bar{p} K P' K$, with P' a *design* positive definite matrix. Under these constraints, we obtain

$$\alpha(z) := u_k$$

$$= -\frac{1}{2} \left(R + \frac{1}{2} g^T(x_k) P g(x_k) \right)^{-1} g^T(x_k) P (f(x_k) - x_{\delta,k+1}). \qquad (4.132)$$

The following theorem establishes the trajectory tracking via inverse optimal control.

Theorem 4.6

Consider that system (2.1) with (4.124) has achieved the SG control goal (4.126) by means of (4.130). Let $V(z_k) = \frac{1}{2} z_k^T P z_k$ be a Lyapunov function candidate with $P = P^T > 0$. Then, the trajectory tracking inverse optimal control law (4.132) renders solution x_k of system (2.1) to be globally asymptotically stable along the desired trajectory $x_{\delta,k}$. Moreover, with $V(x_k) = \frac{1}{2} z_k^T P z_k$ as a CLF and $P = \bar{p} P'$, this control law (4.132) is inverse optimal in the sense that it minimizes the cost functional given by

$$\mathscr{J}(z_k) = \sum_{k=0}^{\infty} \left(l(z_k) + u_k^T R u_k \right), \qquad (4.133)$$

where

$$l(z_k) := -\overline{V} \qquad (4.134)$$

with $\overline{V}$ defined as

$$\overline{V} = V(z_{k+1}) - V(z_k) + \alpha^T(z) R \alpha(z)$$

and $\alpha(z)$ as defined in (4.132).   ∎

Proof

It follows closely the proof given for Theorem 4.5 and hence it is omitted. ■

4.6.1 EXAMPLE

To illustrate the applicability of the proposed methodology, we synthesize a trajectory tracking inverse optimal control law in order to achieve trajectory tracking for a discrete-time second order nonlinear system (unstable for $u_k = 0$) of the form (2.1) with

$$f(x_k) = \begin{bmatrix} 2x_{1,k}\sin(0.5x_{1,k}) + 0.1x_{2,k}^2 \\ 0.1x_{1,k}^2 + 1.8x_{2,k} \end{bmatrix} \qquad (4.135)$$

and

$$g(x_k) = \begin{bmatrix} 0 \\ 2 + 0.1\cos(x_{2,k}) \end{bmatrix}. \qquad (4.136)$$

According to (4.132), the trajectory tracking inverse optimal control law is formulated as

$$u_k = -\frac{1}{2}\left(R + \frac{1}{2}g^T(x_k)Pg(x_k)\right)^{-1} g^T(x_k)P\left(f(x_k) - x_{\delta,k+1}\right),$$

where the positive definite matrix $P_k = p_k P'$ is calculated by the SG algorithm with P' as the identity matrix, that is,

$$
\begin{aligned}
P_k &= p_k P' \\
&= p_k \begin{bmatrix} 0.020 & 0.016 \\ 0.016 & 0.020 \end{bmatrix}
\end{aligned}
$$

and R is a constant matrix

$$R = 0.5.$$

The reference for $x_{2,k}$ is

$$x_{2\delta,k} = 1.5\sin(0.12k)\ rad,$$

and reference $x_{1\delta,k}$ is defined accordingly.

Figure 4.20 presents the trajectory tracking for x_k with initial condition $p_0 = 2.5$ for the SG algorithm, where the solid line $(x_{\delta,k})$ is the reference signal and the dashed line is the evolution of x_k. The control signal is also displayed.

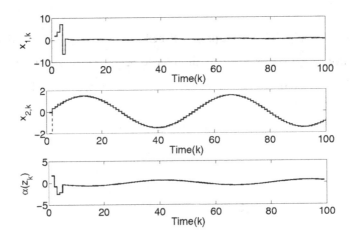

FIGURE 4.20   Tracking performance of x_k.

4.7   TRAJECTORY TRACKING FOR SYSTEMS IN BLOCK-CONTROL FORM

In this section, trajectory tracking is established as a stabilization problem based on an error coordinate transformation for systems in the block-control form. Let us consider that system (2.1) can be presented (possibly after a nonlinear transformation) in the nonlinear block-control form [18] consisting of r blocks as

$$
\begin{aligned}
x_{k+1}^1 &= f^1\left(x_k^1\right) + B^1\left(x_k^1\right) x_k^2 \\
&\vdots \\
x_{k+1}^{r-1} &= f^{r-1}\left(x_k^1, x_k^2, \ldots, x_k^{r-1}\right) \\
&\quad + B^{r-1}\left(x_k^1, x_k^2, \ldots, x_k^{r-1}\right) x_k^r \\
x_{k+1}^r &= f^r\left(x_k\right) + B^r\left(x_k\right) \alpha(x_k),
\end{aligned}
\tag{4.137}
$$

where $x_k \in \mathbb{R}^n$, $x_k = \begin{bmatrix} x_k^{1T} & x_k^{2T} & \cdots & x_k^{rT} \end{bmatrix}^T$; $x^j \in \mathbb{R}^{n_j}$; $j = 1, \ldots, r$; n_j denotes the order

of each r-th block and $n = \sum_{j=1}^r n_j$; input $\alpha(x_k) \in R^m$; $f^j : \mathbb{R}^{n_1 + \cdots + n_j} \to \mathbb{R}^{n_j}$, $B^j :$

$\mathbb{R}^{n_1 + \cdots + n_j} \to \mathbb{R}^{n_j \times n_{j+1}}$, $j = 1, \ldots, r-1$, and $B^r : \mathbb{R}^n \to \mathbb{R}^{n_r \times m}$ are smooth mappings.

Without loss of generality, $x_k = 0$ is an equilibrium point for (4.137). We assume

$f^j(0) = 0$, $rank\{B^j(x_k)\} = n_j \ \forall x_k \neq 0$.

For trajectory tracking of the first block in (4.137), let us define the tracking error

as

$$z_k^1 = x_k^1 - x_{\delta,k}^1,$$
(4.138)

where $x_{\delta,k}^j$ is the desired trajectory.

Once the first new variable is determined (4.138), we take one step ahead

$$z_{k+1}^1 = f^1(x_k^1) + B^1(x_k^1) x_k^2 - x_{\delta,k+1}^1.$$
(4.139)

Equation (4.139) is viewed as a block with state z_k^1 and the state x_k^2 is considered as a

pseudo-control input, where desired dynamics can be imposed, which can be solved

with the anticipation of the desired dynamics for (4.139) as follows:

$$
\begin{aligned}
z_{k+1}^1 &= f^1(x_k^1) + B^1(x_k^1) x_k^2 - x_{\delta,k+1}^1 \\
&= f^1(z_k^1) + B^1(z_k^1) z_k^2.
\end{aligned}
$$
(4.140)

Then, x_k^2 is calculated as

$$x_{\delta,k}^2 = \left(B^1(x_k^1) \right)^{-1} \left(x_{\delta,k+1}^1 - f^1(x_k^1) + f^1(z_k^1) + B^1(z_k^1) z_k^2 \right).$$
(4.141)

Note that the calculated value of state $x_{\delta,k}^2$ in (4.141) is not the real value of such a

state; instead, it represents the desired behavior for x_k^2. To avoid misunderstandings,

the desired value for x_k^2 is referred to as $x_{\delta,k}^2$ in (4.141). Hence, equality (4.140)

is satisfied by substituting the pseudo-control input for x_k^2 in (4.140) as $x_k^2 = x_{\delta,k}^2$,

obtaining $z_{k+1}^1 = f^1(z_k^1) + B^1(z_k^1) z_k^2$. The same procedure is used for each subsequent

block.

Proceeding the same way as for the first block, a second variable in the new coordinates is defined as

$$z_k^2 = x_k^2 - x_{\delta,k}^2.$$

Taking one step ahead in z_k^2 yields

$$
\begin{aligned}
z_{k+1}^2 &= x_{k+1}^2 - x_{\delta,k+1}^2 \\
&= f^2\left(x_k^1, x_k^2\right) + B^2\left(x_k^1, x_k^2\right) x_k^3 - x_{\delta,k+1}^2.
\end{aligned}
$$

The desired dynamics for this block are imposed as

$$
\begin{aligned}
z_{k+1}^2 &= f^2\left(x_k^1, x_k^2\right) + B^2\left(x_k^1, x_k^2\right) x_k^3 - x_{\delta,k+1}^2 \\
&= f^1\left(z_k^1\right) + B^2\left(z_k^1, z_k^2\right) z_k^2.
\end{aligned}
\tag{4.142}
$$

These steps are taken iteratively. At the last step, the known desired variable is $x_{\delta,k}^r$, and the last new variable is defined as

$$z_k^r = x_k^r - x_{\delta,k}^r.$$

As usual, taking one step ahead yields

$$z_{k+1}^r = f^r\left(x_k\right) + B^r\left(x_k\right) \alpha(x_k) - x_{\delta,k+1}^r. \tag{4.143}$$

Finally, the desired dynamics for this last block are imposed as

$$
\begin{aligned}
z_{k+1}^r &= f^r\left(x_k\right) + B^r\left(x_k\right) \alpha(x_k) - x_{\delta,k+1}^r \\
&= f^r\left(z_k\right) + B^r\left(z_k\right) \beta(z_k)
\end{aligned}
\tag{4.144}
$$

which is achieved with

$$\alpha(x_k) = \left(B^r\left(x_k\right)\right)^{-1}\left(x_{\delta,k+1}^r - f^r\left(x_k\right) + f^r\left(z_k\right) + B^r\left(z_k\right) \beta(z_k)\right), \tag{4.145}$$

where $\beta(z_k)$ is proposed as

$$\beta(z_k) = -\frac{1}{2}\left(R + \frac{1}{2}g^T(z_k)P_k g(z_k)\right)^{-1}g^T(z_k)P_k f(z_k). \tag{4.146}$$

Then, system (4.137) can be presented in the new variables $z = \left[z^{1T}z^{2T}\cdots z^{rT}\right]$ as

$$z_{k+1}^1 = f^1(z_k^1) + B^1(z_k^1)z_k^2$$

$$\vdots$$

$$z_{k+1}^{r-1} = f^{r-1}(z_k^1, z_k^2, \ldots, z_k^{r-1}) \tag{4.147}$$

$$+B^{r-1}(z_k^1, z_k^2, \ldots, z_k^{r-1})z_k^r$$

$$z_{k+1}^r = f^r(z_k) + B^r(z_k)\beta(z_k).$$

For system (4.147), we establish the following lemma.

Lemma 4.3

Consider that the equilibrium point $x_k = 0$ of system (4.137) is asymptotically stable by means of the control law (4.112), as established in Theorem 4.5. Then the closed-loop solution of transformed system (4.147) with control law (4.146) is (globally) asymptotically stable along the desired trajectory $x_{\delta,k}$, where $P_k = p_k P$, and p_k is calculated using the proposed SG scheme. Moreover, control law (4.146) is inverse optimal in the sense that it minimizes the cost functional

$$\mathcal{J} = \sum_{k=0}^{\infty}\left(l(z_k) + \beta^T(z_k)R\beta(z_k)\right) \tag{4.148}$$

with

$$l(z_k) = -\overline{V}(z_k) \geq 0. \tag{4.149}$$

■

Proof

Since the equilibrium point $x_k = 0$ for system (4.137) with a control law of the form given in (4.112) is globally asymptotically stable, then the equilibrium point $z_k = 0$ (tracking error) for system (4.147) with control law (4.146) is globally asymptotically stable for the transformed system along the desired trajectory $x_{\delta,k}$.

The minimization of the cost functional is established similarly as in Theorem 4.5, and hence it is omitted. Figure 4.21 displays the proposed transformation and optimality scheme. ■

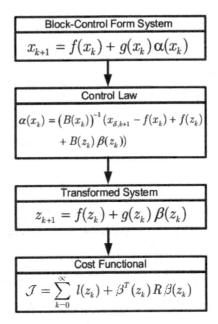

FIGURE 4.21   Transformation scheme and inverse optimal control for the transformed system.

4.7.1 EXAMPLE

In this example, we apply the proposed trajectory tracking inverse optimal control law for a discrete-time second order system (unstable for $u_k = 0$) of the form (2.1)

with

$$f(x_k) = \begin{bmatrix} 1.5x_{1,k} + x_{2,k} \\ x_{1,k} + 2x_{2,k} \end{bmatrix} \qquad (4.150)$$

and

$$g(x_k) = \begin{bmatrix} 0 \\ 1 \end{bmatrix}. \qquad (4.151)$$

In accordance with Section 4.7, control law (4.145) becomes

$$\alpha(x_k) = x^1_{\delta,k+2} - 3.25x_{1,k} - 1.5x_{2,k} + 5.25z_{1,k} + 3.5z_{2,k} + 2\beta(z_k), \qquad (4.152)$$

where $\beta(z_k)$ is given in (4.146), for which $P = p_k P'$, with

$$P' = \begin{bmatrix} 15 & 7 \\ 7 & 15 \end{bmatrix}$$

which is sufficient to ensure asymptotic stability along the desired trajectory $(x_{\delta,k})$; and $R = 0.01$ and p_k are determined by the SG algorithm with initial condition $p_0 = 0.1$. Figure 4.22 presents the trajectory tracking for first block $x_{1,k}$, where the solid line $(x_{\delta,k})$ is the reference signal and the dashed line is the evolution of $x_{1,k}$. The control signal is also displayed. Figure 4.23 displays the SG algorithm time evolution p_k and the respective evaluation of the cost functional $\mathcal{J}$.

4.8 NEURAL INVERSE OPTIMAL CONTROL

Stabilization and trajectory tracking results can be applied to disturbed nonlinear systems, which can be modeled by means of a neural identifier as presented in Chapter 2, obtaining a robust inverse optimal controller. Based on the passivity approach, we propose a neural inverse optimal controller which uses a CLF with a global minimum on the desired trajectory.

    First, the stabilization problem for discrete-time nonlinear systems is discussed.

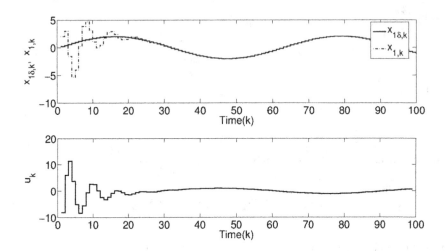

FIGURE 4.22   Tracking performance of x_k for the system in block-control form.

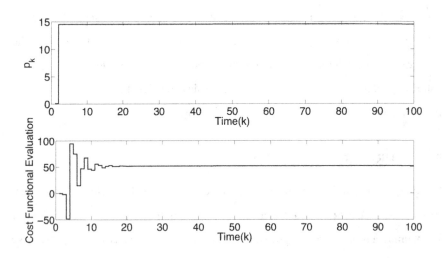

FIGURE 4.23 p_k and $\mathscr{J}$ time evolution for trajectory tracking.

4.8.1 STABILIZATION

As described in Chapter 2, for neural identification of (2.1), a series-parallel neural model (2.40) can be used. Then, for this neural model, stabilization results established in Chapter 2 are applied as follows. Model (2.1) can be represented as a system of the form (2.27)

$$x_{k+1} = f(x_k) + g(x_k) u_k,$$

and if there exists $P = P^T > 0$, this system can be asymptotically stabilized by the inverse optimal control law

$$\alpha(x_k) = -\left(I_m + \frac{1}{2} g^T(x_k) P g(x_k)\right)^{-1} g^T(x_k) P f(x_k).$$

An example illustrates the previously mentioned results.

4.8.1.1 Example

Let us consider a discrete-time second order nonlinear system (unstable for $u_k = 0$) of the form (2.27) with

$$f(x_k) = \begin{bmatrix} 0.5 x_{1,k} \sin(0.5 x_{1,k}) + 0.2 x_{2,k}^2 \\ 0.1 x_{1,k}^2 + 1.8 x_{2,k} \end{bmatrix} \tag{4.153}$$

and

$$g(x_k) = \begin{bmatrix} 0 \\ 2 + 0.1 \cos(x_{2,k}) \end{bmatrix}. \tag{4.154}$$

The phase portrait for this unstable open-loop ($u_k = 0$) system with initial conditions $x_0 = [2, -2]^T$ is displayed in Figure 4.24.

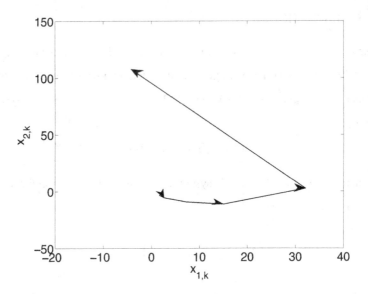

FIGURE 4.24 Unstable system phase portrait.

4.8.1.1.1 Identifier

Let us assume that system (2.27) with (4.153)–(4.154) is unknown. In order to identify this uncertain system, we propose the following series-parallel neural network:

$$
\begin{aligned}
\widehat{x}_{1,k+1} &= w_{11,k} S(x_1) + w_{12,k} (S(x_2))^2 \\
\widehat{x}_{1,k+1} &= w_{21,k} (S(x_1))^2 + w_{22,k} S(x_2) + w_2' u_k,
\end{aligned}
\tag{4.155}
$$

which can be rewritten as $\widehat{x}_{k+1} = f(x_k) + g(x_k) u_k$, where

$$
f(x_k) = \begin{bmatrix} w_{11,k} S(x_1) + w_{12,k} (S(x_2))^2 \\ w_{21,k} (S(x_1))^2 + w_{22,k} S(x_2) \end{bmatrix},
\tag{4.156}
$$

and

$$
g(x_k) = \begin{bmatrix} 0 \\ w_2' \end{bmatrix},
\tag{4.157}
$$

with $w_2' = 0.8$. The initial conditions for the adjustable weights are selected as Gaussian random values, with zero mean and a standard deviation of 0.333; $\eta_1 = \eta_2 = 0.99$,

$P_1 = P_2 = 10I_2$, where I_2 is the 2×2 identity matrix; $Q_1 = Q_2 = 1300I_2$, $R_1 = 1000$, and $R_2 = 4500$.

4.8.1.1.2 Control Synthesis

Then, the inverse optimal control law is formulated as

$$\alpha(x_k) = -\left(1 + \frac{1}{2}g^T(x_k)Pg(x_k)\right)^{-1} g^T(x_k)Pf(x_k), \qquad (4.158)$$

with P as

$$P = \begin{bmatrix} 0.0005 & 0.0319 \\ 0.0319 & 3.2942 \end{bmatrix}.$$

Figure 4.25 presents the stabilization time response for x_k with initial conditions $x_0 = [2, -2]^T$. Initial conditions for RHONN are $\widehat{x}_0 = [0.5, 2]^T$. Figure 4.26 displays the applied inverse optimal control law (4.158), which achieves asymptotic stability; this figure also includes the cost functional values.

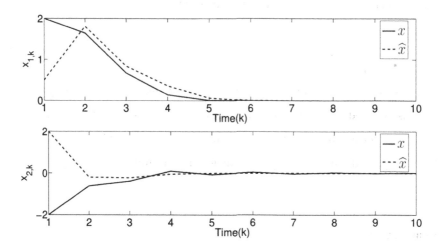

FIGURE 4.25   Stabilized system time response.

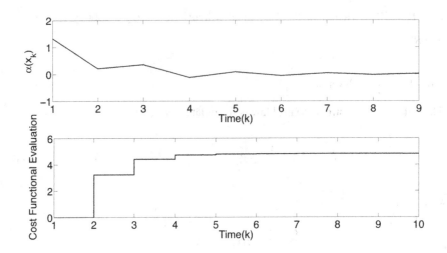

FIGURE 4.26   Control law and cost functional values.

4.8.2   TRAJECTORY TRACKING

The tracking of a desired trajectory, defined in terms of the plant state x_i formulated as (2.40), can be established as the following inequality:

$$\left\| x_{i,\delta} - x_i \right\| \leq \left\| x_i - \widehat{x}_i \right\| + \left\| x_{i,\delta} - \widehat{x}_i \right\|, \tag{4.159}$$

where $\|\cdot\|$ stands for the Euclidean norm, and $x_{i,\delta}$ is the desired trajectory signal, which is assumed smooth and bounded. Inequality (4.159) is valid considering the separation principle for discrete-time nonlinear systems [17], and based on (4.159), the tracking of a desired trajectory can be divided into the following two requirements.

Requirement 4.1

$$\lim_{k \to \infty} \left\| x_i - \widehat{x}_i \right\| \leq \zeta_i \tag{4.160}$$

with ζ_i a small positive constant.

Requirement 4.2

$$\lim_{k \to \infty} \|x_{i,\delta} - \hat{x}_i\| = 0. \tag{4.161}$$

In order to fulfill Requirement 4.1, an on-line neural identifier based on (2.40) is proposed to ensure (4.160) [29], while Requirement 4.2 is guaranteed by a discrete-time controller developed using the inverse optimal control technique. A general control scheme is shown in Figure 4.27. Trajectory tracking is illustrated for the

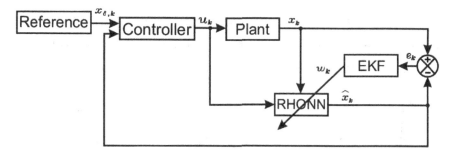

FIGURE 4.27   Control scheme.

neural scheme as follows.

4.8.2.0.1 Example

In order to achieve trajectory tracking for $x_{k+1} = f(x_k) + g(x_k) u_k$ with $f(x_k)$ and $g(x_k)$ as defined by the neural model in (4.156) and (4.157), respectively, then the control law is established as $u_k = -y_k$, for which $y_k = h(x_k, x_{\delta,k+1}) + J(x_k) u_k$, where

$$h(x_k, x_{\delta,k+1}) = g^T(x_k)\overline{P}\left(f(x_k) - x_{\delta,k+1}\right)$$

and

$$J(x_k) = \frac{1}{2} g^T(x_k)\overline{P} g(x_k)$$

with the signal reference for $x_{2,k}$ as

$$x_{2\delta,k} = 2\sin(0.075k) \text{ rad}$$

and reference $x_{1\delta,k}$ is defined accordingly with the system dynamics. Hence, we adjust gain matrix $\overline{P} = K^T P K$ in order to achieve trajectory tracking for $x_k = [x_{1,k}, x_{2,k}]^T$.

Figure 4.28 presents trajectory tracking for x_k with

$$P = \begin{bmatrix} 0.0484 & 0.0387 \\ 0.0387 & 0.0484 \end{bmatrix} ; \ K = \begin{bmatrix} 0.100 & 0.00 \\ 0.00 & 8.25 \end{bmatrix}.$$

Figure 4.29 presents the applied control signal to achieve trajectory tracking; it also displays the cost functional values, which increase because the control law is different from zero in order to achieve trajectory tracking.

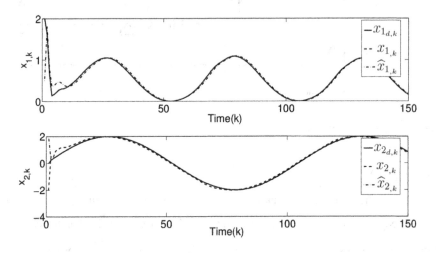

FIGURE 4.28   System time response for trajectory tracking.

4.8.2.1   Application to a Synchronous Generator

In this section, we apply the speed-gradient inverse optimal control technique to a discrete-time neural model identifying a synchronous generator [2]. Figure 4.30 depicts a power synchronous generator. The goal of power system stabilization is to

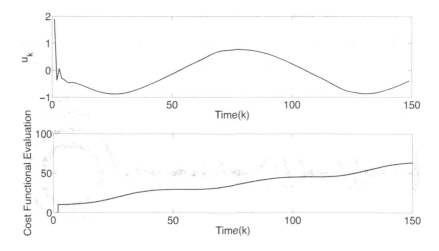

FIGURE 4.29   Control law for trajectory tracking and cost functional values.

produce stable, reliable, and robust electrical energy production and distribution. One task to achieve this goal is to reduce the adverse effects of mechanical oscillations, which can induce premature degradation of mechanical and electrical components as well as dangerous operation of the system (e.g., blackouts) [4].

FIGURE 4.30   Synchronous generator.

4.8.2.1.1 Synchronous Generator Model

We consider a synchronous generator connected through purely reactive transmission lines to the rest of the grid which is represented by an infinite bus (see Figure 4.31).

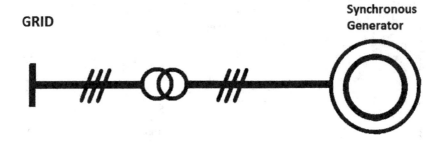

GRID

Synchronous Generator

FIGURE 4.31   Synchronous generator connected to infinite bus.

The discrete-time model of the synchronous generator is given by [15]

$$
\begin{aligned}
x_{1,k+1} &= f^1\left(\bar{x}_k^1\right) + \tau x_{2,k} \\
x_{2,k+1} &= f^2\left(\bar{x}_k^2\right) + \tau m_2 x_{3,k} \\
x_{3,k+1} &= f^3\left(\bar{x}_k^3\right) + \tau m_6 u_k
\end{aligned}
\tag{4.162}
$$

with

$$
f^1\left(\bar{x}_k^1\right) = x_{1,k}
$$

$$
f^2\left(\bar{x}_k^2\right) = x_{2,k} + \tau\left[m_1 + \left(m_2 E_q^{'*} + m_3 \cos\left(\tilde{x}_1\right)\right)\sin\left(\tilde{x}_1\right)\right]
$$

$$
f^3\left(\bar{x}_k^3\right) = x_{3,k} + \tau\left[m_4\left(x_{3,k} + E_q^{'*}\right) + m_5 \cos\left(\tilde{x}_1\right) + m_6 E_{fd}^*\right]
$$

and

$$
\tilde{x}_1 = x_{1,k} + \delta^*, \quad m_1 = \frac{T_m}{M}, \quad m_2 = \frac{-V}{MX_d'}, \quad m_3 = \frac{V^2}{M}\left(\frac{1}{X_d'} - \frac{1}{X_q}\right),
$$

$$
m_4 = -\frac{X_d}{T_{do}'X_d'}, \quad m_5 = -\left(\frac{X_d' - X_d}{T_{do}'X_d'}\right)V, \quad m_6 = \frac{1}{T_{do}'},
$$

with

$$x_1 \quad := \quad \Delta\delta = \delta - \delta^*$$

$$x_2 \quad := \quad \Delta\omega = \omega - \omega^* \qquad (4.163)$$

$$x_3 \quad := \quad \Delta E'_q = E'_q - E'^*_q,$$

where δ is the generator rotor angle referred to the infinite bus (also called the power angle), ω is the rotor angular speed, and E'_q is the stator voltage which is proportional to flux linkages; τ is the sampling time, M is the per-unit inertia constant, T_m is the constant mechanical power supplied by the turbine, and T'_{do} is the transient open circuit time constant. $X_d = x_d + x_L$ is the augmented reactance, where x_d is the direct axis reactance and x_L is the line reactance, X'_d is the transient augmented reactance, X_q is the quadrature axis reactance, and V is the infinite bus voltage, which is fixed. The generated power P_g and the stator equivalent voltage E_{fd} are given as

$$P_g \quad = \quad \frac{1}{X'_d} E'_q V \sin(\delta) + \frac{1}{2}\left(\frac{1}{X_q} - \frac{1}{X'_d}\right) V^2 \sin(2\delta)$$

$$E_{fd} \quad = \quad \frac{\omega_s M_f}{\sqrt{2}R_f} v_f,$$

respectively, where v_f is the scaled field excitation voltage, M_f is the mutual inductance between stator coils, R_f is the field resistance, and ω_s is the synchronous speed. As in [15], we only consider the case where the dynamics of the damper windings are neglected.

Through this work, the analysis and design are done around an operation point $\left(\delta^*, \omega^*, E'^*_q\right)$, which is obtained for a stator field equivalent voltage $E^*_{fd} = 1.1773$ as proposed in [15], for which $\delta^* = 0.870204$, $\omega^* = 1$, and $E'^*_q = 0.822213$. The sampling time is selected as $\tau = 0.01$.

The parameters of the plant model used for simulation are given in Table 4.1.

TABLE 4.1

Model Parameters (per Unit)

PARAMETER	VALUE		PARAMETER	VALUE
T_m	1		X_q	0.9
M	0.033		X_d	0.9
ω_s	0.25		X'_d	0.3
T'_{do}	0.033		V	1.0

4.8.2.1.2 Neural Identification for the Synchronous Generator

In order to identify the discrete-time synchronous generator model (4.162), we propose a RHONN as follows:

$$
\begin{aligned}
\widehat{x}_{1,k+1} &= w_{11,k}S\left(x_{1,k}\right) + w_{12,k}S\left(x_{2,k}\right) \\
\widehat{x}_{2,k+1} &= w_{21,k}S\left(x_{1,k}\right)^6 + w_{22,k}S\left(x_{2,k}\right)^2 + w_{23,k}S\left(x_{3,k}\right) \qquad (4.164) \\
\widehat{x}_{3,k+1} &= w_{31,k}S\left(x_{1,k}\right)^2 + w_{32,k}S\left(x_{2,k}\right) + w_{33,k}S\left(x_{3,k}\right)^2 + w_{34}u_k,
\end{aligned}
$$

where $\widehat{x}_i$ estimates x_i $(i = 1, 2, 3)$, and w_{34} is a fixed parameter in order to ensure the controllability of the neural identifier [25], which is selected as $w_{34} = 0.5$. System (4.165) can be rewritten as

$$
x_{k+1} = f(x_k) + g(x_k)u_k
$$

with

$$
f(x_k) = \begin{bmatrix} w_{11,k}S\left(x_{1,k}\right) + w_{12,k}S\left(x_{2,k}\right) \\ w_{21,k}S\left(x_{1,k}\right)^6 + w_{22,k}S\left(x_{2,k}\right)^2 + w_{23,k}S\left(x_{3,k}\right) \\ w_{31,k}S\left(x_{1,k}\right)^2 + w_{32,k}S\left(x_{2,k}\right) + w_{33,k}S\left(x_{3,k}\right)^2 \end{bmatrix} \qquad (4.165)
$$

and

$$
g(x_k) = \begin{bmatrix} 0 \\ 0 \\ w_{34} \end{bmatrix}.
$$
(4.166)

The training is performed on-line, using a series-parallel configuration, with the EKF learning algorithm. All the NN states are initialized in a random way as well as the weights vectors. It is important to remark that the initial conditions of the plant are completely different from the initial conditions of the NN. The covariance matrices are initialized as diagonals, and the nonzero elements are $P_1(0) = P_2(0) = 10,000$; $Q_1(0) = Q_2(0) = 5000$ and $R_1(0) = R_2(0) = 10,000$, respectively.

4.8.2.1.3 Control Synthesis

For system (2.1), with $f(x_k)$ and $g(x_k)$ as defined in (4.165) and (4.166), respectively, the proposed inverse optimal control law (4.101) is formulated as

$$
u_k^* = -\frac{1}{2}\left(R + \frac{1}{2}g^T(x_k)P_k g(x_k)\right)^{-1} g^T(x_k)P_k f(x_k),
$$

where R is a constant matrix selected as $R = 1$, and P_k is defined as $P_k = p_k P'$, for which p_k is calculated by the SG algorithm, and P' is a symmetric and positive definite matrix selected as

$$
P' = \begin{bmatrix} 1.5 & 1.0 & 0.5 \\ 1.0 & 1.5 & 1.0 \\ 0.5 & 1.0 & 1.5 \end{bmatrix}.
$$
(4.167)

4.8.2.1.4 Simulation Results

Initially, results for the inverse optimal controller based on the plant model are portrayed. Figure 4.32 displays the solutions of $\left[\delta, \omega, E_q'\right]^T$, which are based on the systems (4.162) and (4.163), with initial conditions $[0.77, 0.10, 0.85]^T$, respectively. Additionally, in order to illustrate the robustness of the proposed controller, the simulation stages are indicated as follows:

Stage 1: Nominal parameters, as given in Table 4.1, are used at the beginning of the simulation.

Stage 2: A short circuit fault occurs at 1.5 seconds, which is equivalent to changing the augmented reactance X_d from $X_d = 0.9$ to $X_d = 0.1$.

Stage 3: The short circuit fault is removed at 1.6 seconds.

Stage 4: A disturbance in the mechanical power is carried out by changing T_m from $T_m = 1$ to $T_m = 1.2$ at 3.5 seconds.

Stage 5: The disturbance is removed at 3.6 seconds.

Stage 6: The system is in a post-fault and post-disturbance state.

Figure 4.33 includes the applied inverse optimal control law, the time evolution for parameter p_k, and the cost functional evaluation.

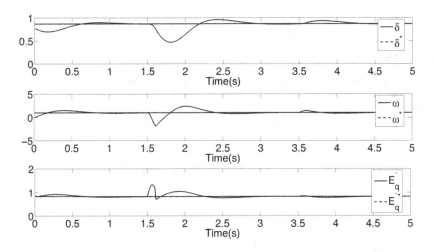

FIGURE 4.32   Time evolution for x_k.

Finally, the application of the proposed inverse optimal neural controller based on the neural identifier is presented. Figure 4.34 displays the solution of $\left[\delta, \omega, E_q'\right]^T$, for the neural identifier (4.165), with the initial conditions given as $[0.77, 0.10, 0.85]^T$, respectively. In order to illustrate the robustness of the proposed neural controller, the same simulation stages, as explained above, were carried out for this neural scheme. Figure 4.35 includes the applied inverse optimal neural control law, the time evolution for parameter p_k, and the cost functional evaluation. The control goal is to guarantee that all the state variable values stay at their equilibrium point (regulation),

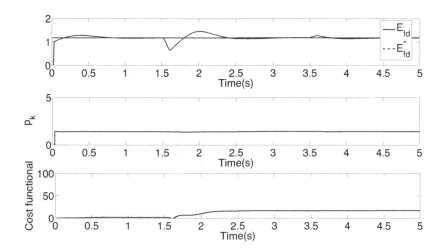

FIGURE 4.33   Time evolution for control law u_k, parameter p_k, and the cost functional.

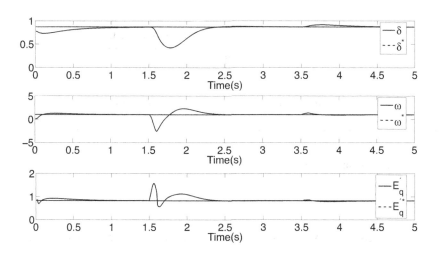

FIGURE 4.34   Time evolution for x_k using the neural identifier.

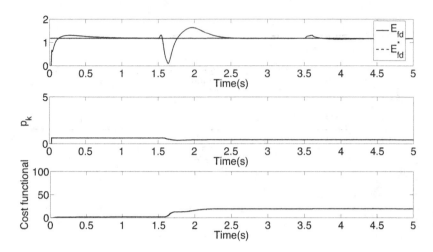

FIGURE 4.35   Time evolution for control law u_k, parameter p_k, and the cost functional using the neural identifier.

TABLE 4.2

Comparison of the Regulation Error with Short Circuit

	SG-IOC	SG-IONC
Mean value	0.9018	0.8980
Standard deviation	0.1081	0.0988

as illustrated by the simulations.

4.8.2.2 Comparison

In order to compare the proposed control schemes, Table 4.2 is included, and is described as follows. The controllers used in this comparison are: 1) speed-gradient inverse optimal control (SG-IOC) and 2) speed-gradient inverse optimal neural control (SG-IONC).

It is important to note that for the SG-IOC, it is required to know the exact plant model (structure and parameters) for the controller synthesis, which in many cases is a disadvantage [26].

4.9 BLOCK-CONTROL FORM: A NONLINEAR SYSTEMS PARTICU-
LAR CLASS

In this section, we develop a neural inverse optimal control scheme for systems which have a special state representation referred as the block-control (BC) form [19]. A block transformation for a neural identifier is applied in order to obtain an error system on the desired reference, and then a neural inverse optimal stabilization control law for the error resulting system is synthesized.

4.9.1 BLOCK TRANSFORMATION

Let us consider that system (2.1), under an appropriate nonsingular transformation, can be rewritten as the following BC form with r blocks:

$$
\begin{aligned}
x_{i_1,k+1} &= f_{i_1}(x_{i_1,k}) + B_{i_1}(x_{i_1,k})x_{i_1,k} + \Gamma_{i_1,k} \\
x_{i_2,k+1} &= f_{i_2}(x_{i_1,k}, x_{i_2,k}) + B_{i_2}(x_{i_1,k}, x_{i_2,k})xi_3, k + \Gamma_{i_2,k} \\
&\vdots \\
x_{i_r,k+1} &= f_{i_r}(x_k) + B_{i_r}(x_k)u_k + \Gamma_{i_r,k},
\end{aligned}
\tag{4.168}
$$

where $x_k \in \mathbb{R}^n$ is the system state, $x_k = \begin{bmatrix} x_{i_1,k}^T, & x_{i_2,k}^T, & \dots, & x_{i_r,k}^T \end{bmatrix}^T$; $i = 1,\dots,n_r$; $u_k \in \mathbb{R}^{m_r}$ is the input vector. We assume that f_{i_j}, B_{i_j}, and Γ_{i_j} are smooth functions, $j = 1,\dots,r$, $f_{i_j}(0) = 0$, and $rank\{B_{i_j}(\chi_k)\} = m_j \ \forall \chi_k \neq 0$. The unmatched and matched disturbance terms are represented by Γ_i. The whole system order is $n = \sum_{j=1}^{r} n_j$.

To identify (4.168), we propose a neural network with the same BC structure, consisting of r blocks as follows:

$$
\begin{aligned}
\hat{x}_{1,k+1} &= W_{i_1,k}\,\rho_{i_1}(x_{i_1,k}) + W'_{i_1}\,x_{i_2,k} \\
\hat{x}_{i_2,k+1} &= W_{i_2,k}\,\rho_{i_2}(x_{i_1,k},x_{i_2,k}) + W'_{i_2}\,x_{i_3,k} \\
&\;\;\vdots \\
\hat{x}_{i_r,k+1} &= W_{i_r,k}\,\rho_{i_r}(x_{i_1,k},\dots,x_{i_r,k}) + W'_{i_r}\,u_k,
\end{aligned}
\tag{4.169}
$$

where $\hat{x}_k = \begin{bmatrix} \hat{x}_{i_1}^T, & \hat{x}_{i_2}^T, & \dots, & \hat{x}_{i_r}^T \end{bmatrix}^T \in \mathbb{R}^n$, $\hat{x}_{i_r} \in \mathbb{R}^{n_r}$ denotes the i-th neu-

ron system state corresponding to the r-th block; $i = 1,\dots,n_r$; $W_{i_1,k} = \begin{bmatrix} w_{1_1,k}^T, & w_{2_1,k}^T, & \dots, & w_{n_{r1},k}^T \end{bmatrix}^T$ is the on-line adjustable weight matrix, and $W'_{i_r} = \begin{bmatrix} w_{1_1}'^T, & w_{2_1}'^T, & \dots, & w_{n_{r1}}'^T \end{bmatrix}^T$ is the fixed weight matrix; n_r denotes the order for

each r-th block, and the whole system order becomes $n = \sum\limits_{j=1}^{r} n_j$.

First, we define the tracking error as

$$
z_{i_1,k} = \hat{x}_{i_1,k} - x_{i_1\delta,k},
\tag{4.170}
$$

where $x_{i_1\delta,k}$ is the desired trajectory signal.

Once the first new variable (4.170) is defined, one step ahead is taken as

$$
z_{i_1,k+1} = W_{i_1,k}\,\rho_{i_1}(x_{i_1,k}) + W'_{i_1}\,x_{i_2,k} - x_{i_1\delta,k+1}.
\tag{4.171}
$$

Equation (4.171) is viewed as a block with state $z_{i_1,k}$ and the state $x_{i_2,k}$ is considered as a pseudo-control input, where desired dynamics can be imposed, which can be solved with the anticipation of the desired dynamics for this block as follows:

$$
z_{i_1,k+1} = W_{i_1,k}\,\rho_{i_1}(x_{i_1,k}) + W'_{i_1}\,x_{i_2,k} - x_{i_1\delta,k+1} = K_{i_1}\,z_{i_1,k},
\tag{4.172}
$$

where $K_{i_1} = diag\{k_{11},\dots,k_{n_11}\}$ with $|k_{q1}| < 1$, $q = 1,\dots,n_1$, in order to assure stability for block (4.172). From (4.172), $x_{i_2,k}$ is calculated as

$$x_{i_2\delta,k} = \left(W_{i_1}'\right)^{-1}\left(-W_{i_1,k}\rho_{i_1}\left(x_{i_1,k}\right)+x_{i_1\delta,k+1}+K_{i_1}z_{i_1,k}\right).$$

Note that the calculated value for state $x_{i_2\delta,k}$ in (4.173) is not the real value for such a state; instead, it represents the desired behavior for $x_{i_2,k}$. Hence, to avoid confusion this desired value of $x_{i_2,k}$ is referred to as $x_{i_2\delta,k}$ in (4.173).

Proceeding along the same way as for the first block, a second variable in the new coordinates is defined as

$$z_{i_2,k} = \widehat{x}_{i_2,k} - x_{i_2\delta,k}.$$

Taking one step ahead in $z_{i_2,k}$ yields

$$\begin{aligned}
z_{i_2,k+1} &= \widehat{x}_{i_2,k+1} - x_{i_2\delta,k+1} \\
&= W_{i_2,k}\rho_{i_2}\left(x_{i_1,k},x_{i_2,k}\right)+W_{i_2}'x_{i_3,k}-x_{i_2\delta,k+1}.
\end{aligned}$$

The desired dynamics for this block are imposed as

$$\begin{aligned}
z_{i_2,k+1} &= W_{i_2,k}\rho_{i_2}\left(x_{i_1,k},x_{i_2,k}\right)+W_{i_2}'x_{i_3,k}-x_{i_2\delta,k+1} \\
&= K_{i_2}z_{i_2,k}, && (4.173)
\end{aligned}$$

where $K_{i_2}=diag\{k_{12},\dots,k_{n_22}\}$ with $|k_{q2}|<1, q=1,\dots,n_2$.

These steps are taken iteratively. At the last step, the known desired variable is $x_{r\delta,k}^r$, and the last new variable is defined as

$$z_{i_r,k} = \widehat{x}_{i_r,k} - x_{i_r\delta,k}.$$

As usual, taking one step ahead yields

$$z_{i_r,k+1} = W_{i_r,k}\rho_{i_r}\left(x_{i_1,k},\dots,x_{i_r,k}\right)+W_{i_r}'u_k-x_{i_r\delta,k+1}. \qquad (4.174)$$

System (4.169) can be represented in the new variables as

$$
\begin{aligned}
z_{i_1,k+1} &= K_1 z_{i_1,k} + W'_{i_1} z_{i_2,k} \\
z_{i_2,k+1} &= K_{i_2} z_{i_2,k} + W'_{i_2} z_{i_3,k} \\
&\;\;\vdots \\
z_{i_r,k+1} &= W_{i_r,k}\, \rho_{i_r}(x_{i_1,k}, \dots, x_{i_r,k}) - x_{i_r\delta,k+1} + W'_{i_r} u_k.
\end{aligned}
\tag{4.175}
$$

4.9.2 BLOCK INVERSE OPTIMAL CONTROL

In order to achieve trajectory tracking along $x_{\delta,k}$, the transformed system (4.175) is stabilized at its origin. System (4.175) can be presented in a general form as

$$
z_{k+1} = f(z_k) + g(z_k)\, u_k,
\tag{4.176}
$$

where $z_k = \begin{bmatrix} z_{i_1,k}^T, & z_{i_2,k}^T, & \cdots, & z_{i_r,k}^T \end{bmatrix}^T$.

Then, in order to achieve stabilization for (4.176) along the desired trajectory, we apply the inverse optimal control law as

$$
u_k = \alpha(z_k) = -\left[I_m + J(z_k) \right]^{-1} h(z_k)
\tag{4.177}
$$

with

$$
h(z_k) = g(z_k)^T P f(z_k)
\tag{4.178}
$$

and

$$
J(z_k) = \frac{1}{2} g(z_k)^T P g(z_k).
\tag{4.179}
$$

4.10 CONCLUSIONS

This chapter has established the inverse optimal control technique for a class of discrete-time nonlinear systems. To avoid the solution of the HJB equation, we propose a discrete-time CLF in a quadratic form, which depends on a fixed parameter to achieve stabilization and trajectory tracking. Based on this CLF, the inverse optimal control strategy is synthesized. The proposed approach is extended to discrete-time disturbed nonlinear systems, which results in a robust inverse optimal control, avoiding the solution of the HJI equation. Also, a controller for a class of positive systems is proposed.

On the other hand, a CLF in a quadratic form adjusted by means of the speed-gradient algorithm is proposed. These results are extended to establish a speed-gradient inverse optimal control for trajectory tracking. Additionally, an inverse optimal control scheme is presented for nonlinear systems in block-control form. Finally, on the basis of a neural model of the plant, identified by a RHONN, a neural optimal controller is developed.

REFERENCES

1. A. Al-Tamimi and F. L. Lewis. Discrete-time nonlinear HJB solution using approximate dynamic programming: Convergence proof. *Systems, Man, Cybernetics—Part B, IEEE Transactions on* , 38(4):943–949, 2008.

2. A. Y. Alanis, E. N. Sanchez, and A. G. Loukianov. Discrete-time backstepping synchronous generator stabilization using a neural observer. In *Proceedings of the 17th IFAC World Congress*, pages 15897–15902, Seoul, Korea, 2008.

3. B. D. O. Anderson and J. B. Moore. *Optimal Control: Linear Quadratic Methods*. Prentice-Hall, Englewood Cliffs, NJ, USA, 1990.

4. M. A. Arjona, R. Escarela-Perez, G. Espinosa-Perez, and J. Alvarez-Ramirez. Validity testing of third-order nonlinear models for synchronous generators. *Electric Power Systems Research*, (79):953–958, 2009.

5. G. Escobar, R. Ortega, H. Sira-Ramirez, J. P. Vilian, and I. Zein. An experimental comparison of several non-linear controllers for power converters. *IEEE Control Systems Magazine*, 19(1):66–82, 1999.

6. A. L. Fradkov and A. Y. Pogromsky. *Introduction to Control of Oscillations and Chaos*. World Scientific Publishing Co., Singapore, 1998.

7. R. A. Freeman and P. V. Kokotović. *Robust Nonlinear Control Design: State-Space and Lyapunov Techniques*. Birkhauser Boston Inc., Cambridge, MA, USA,

1996.

8. R. A. Freeman and J. A. Primbs. Control Lyapunov functions: New ideas from an old source. In *Proceedings of the 35th IEEE Conference on Decision and Control*, pages 3926–3931, Kobe, Japan, Dec 1996.

9. J. W. Grizzle, M. D. Benedetto, and L. Lamnabhi-Lagarrigue. Necessary conditions for asymptotic tracking in nonlinear systems. *Automatic Control, IEEE Transactions on*, 39(9):1782–1794, 1994.

10. A. Isidori. *Nonlinear Control Systems*. Springer-Verlag, London, UK, 1997.

11. J. G. Kassakian, M. F. Schlecht, and G. C. Verghese. *Principles of Power Electronics*. Adison Wesley, MA, USA, 1991.

12. H. K. Khalil. *Nonlinear Systems*. Prentice-Hall, Upper Saddle River, NJ, USA, 1996.

13. D. E. Kirk. *Optimal Control Theory: An Introduction*. Prentice-Hall, Englewood Cliffs, NJ, USA, 1970.

14. M. Krstić and H. Deng. *Stabilization of Nonlinear Uncertain Systems*. Springer-Verlag, Berlin, Germany, 1998.

15. J. De Leon-Morales, O. Huerta-Guevara, L. Dugard, and J. M. Dion. Discrete-time nonlinear control scheme for synchronous generator. In *Proceedings of the 42nd Conference on Decision and Control*, pages 5897–5902, Maui, Hawaii, USA, Dec 2003.

16. F. L. Lewis and V. L. Syrmos. *Optimal Control*. John Wiley & Sons, New York, USA, 1995.

17. W. Lin and C. I. Byrnes. Design of discrete-time nonlinear control systems via smooth feedback. *Automatic Control, IEEE Transactions on*, 39(11):2340–2346, 1994.

18. A. G. Loukianov. Nonlinear block control with sliding modes. *Automation and Remote Control*, 57(7):916–933, 1998.

19. A. G. Loukianov and V. I. Utkin. Methods of reducing equations for dynamic systems to a regular form. *Automation and Remote Control*, 42(4):413–420, 1981.

20. Y. Maeda. Euler's discretization revisited. In *Proceedings of the Japan Academy*, volume 71, pages 58–61, 1995.

21. M. Margaliot and G. Langholz. Some nonlinear optimal control problems with closed-form solutions. *International Journal of Robust and Nonlinear Control*, 11(14):1365–1374, 2001.

22. R. Marino and P. Tomei. *Nonlinear Control Design: Geometric, Adaptive and Robust*. Prentice Hall, Hertfordshire, UK, 1996.

23. P. J. Moylan. Implications of passivity in a class of nonlinear systems. *IEEE Transactions on Automatic Control*, 19(4):373–381, 1974.

24. E. M. Navarro-López. Local feedback passivation of nonlinear discrete-time systems through the speed-gradient algorithm. *Automatica*, 43(7):1302–1306, 2007.

25. F. Ornelas-Tellez, A. G. Loukianov, E. N. Sanchez, and E. J. Bayro-Corrochano. Decentralized neural identification and control for uncertain nonlinear systems: Application to planar robot. *Journal of the Franklin Institute*, 347(6):1015–1034, 2010.

26. F. Ornelas-Tellez, E. N. Sanchez, and A. G. Loukianov. Discrete-time robust inverse optimal control for a class of nonlinear systems. In *Proceedings of the 18th IFAC World Congress*, pages 8595–8600, Milano, Italy, 2011.

27. R. Ortega, A. Loría, P. J. Nicklasson, and H. Sira-Ramirez. *Passivity-based Control of Euler-Lagrange Systems: Mechanical, Electrical and Electromechanical Applications*. Springer-Verlag, Berlin, Germany, 1998.

28. J. A. Primbs, V. Nevistic, and J. C. Doyle. Nonlinear optimal control: A control Lyapunov function and receding horizon perspective. *Asian Journal of Control*, 1:14–24, 1999.

29. E. N. Sanchez, A. Y. Alanis, and A. G. Loukianov. *Discrete-time High Order Neural Control*. Springer-Verlag, Berlin, Germany, 2008.

30. E. N. Sanchez and J. P. Perez. Input-to-state stability (iss) analysis for dynamic neural networks. *IEEE Transactions on Circuits and Systems—I: Fundamental Theory and Applications*, 46(11):1395–1398, 1999.

31. R. Sepulchre, M. Jankovic, and P. V. Kokotović. *Constructive Nonlinear Control*. Springer-Verlag, Berlin, Germany, 1997.

Section II

Real-Time Applications

5 Induction Motors

Relevance of the proposed control schemes is illustrated by means of application to induction motors, using rapid control prototyping (RCP); this kind of motors is widely used in industrial applications due to their reliability, simpler construction and reduced cost with respect, for example, to D.C. motors. The neural controller has three components: system identification, trajectory tracking and state estimation, which are solved independently. The neural identifier is based on a RHONN, trained with an EKF. The controllers are used to force the system to track a desired trajectory and to reject undesired disturbances. A super-twisting observer is implemented to estimate the rotor magnetic fluxes. Experimental results illustrate the adequate performance and effectiveness of the control schemes and the RCP system.

Induction motors are one of the most preferred actuators for industrial applications due to their reliability, ruggedness and relatively low maintenance cost. There are electrical vehicles using induction motors, which do not pollute the environment by burning fuel and for which speed and torque control is fundamental [6]. Induction motor control constitutes a challenge since their behavior is nonlinear and some of their variables are not measurable.

A classical technique for induction motors is the field oriented control (FOC), introduced in [5]. More recently, various nonlinear control approaches have been applied to induction motors improving their performance, e.g., sliding mode [15, 10, 21], adaptive input-output linearization [14] and backstepping [18]. All these approaches use continuous time models. The discrete-time case is discussed in [13, 12].

For many nonlinear systems, it is often difficult to obtain their accurate and faithful mathematical models, due to external disturbances, uncertain parameters, and unmodeled dynamics. Therefore, system identification becomes a relevant issue and even necessary before system control can be considered, not only for understanding and predicting the behavior of the system but also to obtain an effective control law. Identification consists of the selection of an appropriate identification model and adjusting its parameters according to an adaptive law [2]. In particular, the use of RHONN training with an EKF algorithm for modeling and control has proven to be reliable and practical for many applications [1, 3, 17].

Rapid control prototyping (RCP) is one of the most important technologies for speeding up product development time. The key element of the RCP is automatic code generation, which eliminates error-prone hand coding procedures, thus making it possible for engineers to focus on control system synthesis, implementation and evaluation. Texas Instruments TMS320F28069M is a low-cost microcontroller for real-time control applications. It belongs to the C2000 MCU family with a 32-bit C28x DSP core and a real-time control accelerator (CLA) to increase the bandwidth of the C28x core, up to 240 MIPS of total performance.

5.1 NEURAL IDENTIFIER

In order to obtain a discrete-time neural model for the induction motor, it is assumed that all states are measured. The RHONN identifier is proposed as:

$$\hat{x}_{1,k+1} = w_{11,k}S(\theta_k) + w_{12,k}S(w_k)$$

$$\hat{x}_{2,k+1} = w_{21,k}S(w_k) - w'_{21}S(\phi_{\beta,k})i_{\alpha,k} + w'_{22}S(\phi_{\alpha,k})i_{\beta,k}$$

$$\hat{x}_{3,k+1} = w_{31,k}S(\Phi_{m,k}) + w'_{31}S(\phi_{\alpha,k})i_{\alpha,k} + w'_{32}S(\phi_{\beta,k})i_{\beta,k}$$

$$\hat{x}_{4,k+1} = w_{41,k}S(\phi_{\alpha,k})\omega_k + w_{42,k}S(\phi_{\beta,k})\omega_k + w_{43,k}S(i_{\alpha,k}) + w'_{41}u_{\alpha,k}$$

$$\hat{x}_{5,k+1} = w_{51,k}S(\phi_{\alpha,k})\omega_k + w_{52,k}S(\phi_{\beta,k})\omega_k + w_{53,k}S(i_{\beta,k}) + w'_{51}u_{\beta,k},$$

where $\hat{x}_{1,k}$ estimates the rotor angular position $\theta(k)$; $\hat{x}_{2,k}$ estimates the rotor angular velocity $\omega(k)$; $\hat{x}_{3,k}$ estimates the square rotor flux magnitude Φ_m; $\hat{x}_{4,k}$ and $\hat{x}_{5,k}$ estimates the stator currents $i_{\alpha,k}$ and $i_{\beta,k}$, respectively. The training is performed with all neural network states $\hat{x}_{i,k}$ initialized randomly as well as the weights $w_{ij,k}$ and with $w'_{21} = w'_{22} = 0.1$, $w'_{31} = w'_{32} = 0.00092$ and $w'_{41} = w'_{51} = 0.0092$. The covariances matrices are initialized as diagonals, and non-zero elements. The neural network structure (5.1) is determined heuristically in order to minimize the state estimation error.

5.2 DISCRETE-TIME SUPER-TWISTING OBSERVER

A discrete-time super-twisting observer structure proposed in [16] is used for rotor flux estimations. The flux observer is defined as

$$
\begin{aligned}
\hat{i}_{\alpha,k+1} =& \hat{i}_{\alpha,k} + \delta(\alpha\beta\hat{\phi}_{\alpha,k} + n_p\beta\omega_k\hat{\phi}_{\beta,k} - \gamma\hat{i}_{\alpha,k}) \\
& + \frac{\delta}{\sigma}u_{\alpha,k} + c_1\delta\sqrt{|\tilde{i}_{\alpha,k}|}\mathrm{sign}\tilde{i}_{\alpha,k} + c_2\delta\tilde{i}_{\alpha,k} \\
\hat{i}_{\beta,k+1} =& \hat{i}_{\beta,k} + \delta(\alpha\beta\hat{\phi}_{\beta,k} - n_p\beta\omega_k\hat{\phi}_{\alpha,k} - \gamma\hat{i}_{\beta,k}) \\
& + \frac{\delta}{\sigma}u_{\beta,k} + c_3\delta\sqrt{|\tilde{i}_{\beta,k}|}\mathrm{sign}\tilde{i}_{\beta,k} + c_4\delta\tilde{i}_{\beta,k} \\
\hat{\phi}_{\alpha,k+1} =& \overline{\vartheta}_{1,k} + c_5\delta\mathrm{sign}\tilde{i}_{\alpha,k} + c_6\delta\tilde{i}_{\alpha,k} \\
\hat{\phi}_{\beta,k+1} =& \overline{\vartheta}_{2,k} + c_7\delta\mathrm{sign}\tilde{i}_{\beta,k} + c_8\delta\tilde{i}_{\beta,k},
\end{aligned}
\tag{5.1}
$$

where $\hat{\phi}_{\alpha,k}$, $\hat{\phi}_{\beta,k}$, $\hat{i}_{\alpha,k}$ and $\hat{i}_{\beta,k}$ estimate the variables $\phi_{\alpha,k}$, $\phi_{\beta,k}$, $i_{\alpha,k}$ and $i_{\beta,k}$ respectively, $\tilde{i}_{\alpha,k} = \hat{i}_{\alpha,k} - i_{\alpha,k}$ $\tilde{i}_{\beta,k} = \hat{i}_{\beta,k} - i_{\beta,k}$ are the currents observation errors and

$$
\begin{aligned}
\overline{\vartheta}_{1,k} =& \overline{\rho}_{1,k}\cos\overline{\Delta\theta}_k - \overline{\rho}_{2,k}\sin\overline{\Delta\theta}_k \\
\overline{\vartheta}_{2,k} =& \overline{\rho}_{1,k}\sin\overline{\Delta\theta}_k + \overline{\rho}_{2,k}\cos\overline{\Delta\theta}_k \\
\overline{\rho}_{1,k} =& a_0\hat{\phi}_{\alpha,k} + a_3 i_{\alpha,k} \\
\overline{\rho}_{2,k} =& a_0\hat{\phi}_{\beta,k} + a_3 i_{\beta,k} \\
\overline{\Delta\theta}_k =& n_p\delta\omega_k + n_p a_1(i_{\beta,k}\hat{\phi}_{\alpha,k} - i_{\alpha,k}\hat{\phi}_{\beta,k}).
\end{aligned}
$$

The dynamics for the observation errors, defined as $\tilde{x}_k = \hat{x}_k - x_k$, have the form

$$
\begin{aligned}
\tilde{i}_{\alpha,k+1} &= \tilde{i}_{\alpha,k} + c_1\delta\sqrt{|\tilde{i}_{\alpha,k}|}\mathrm{sign}\tilde{i}_{\alpha,k} + c_2\delta\tilde{i}_{\alpha,k} + \Lambda_{1,k} \\
\tilde{i}_{\beta,k+1} &= \tilde{i}_{\beta,k} + c_3\delta\sqrt{|\tilde{i}_{\beta,k}|}\mathrm{sign}\tilde{i}_{\beta,k} + c_4\delta\tilde{i}_{\beta,k} + \Lambda_{2,k} \qquad (5.2) \\
\tilde{\phi}_{\alpha,k+1} &= a_0\tilde{\phi}_{\alpha,k} + c_5\delta\mathrm{sign}\tilde{i}_{\alpha,k} + c_6\delta\tilde{i}_{\alpha,k} \\
\tilde{\phi}_{\beta,k+1} &= a_0\tilde{\phi}_{\beta,k} + c_7\delta\mathrm{sign}\tilde{i}_{\beta,k} + c_8\delta\tilde{i}_{\beta,k},
\end{aligned}
$$

with,

$$
\begin{aligned}
\Lambda_{1,k} &= \delta(\alpha\beta\tilde{\phi}_{\alpha,k} + n_p\beta\omega_k\tilde{\phi}_{\beta,k} - \gamma\tilde{i}_{\alpha,k}) \\
\Lambda_{2,k} &= \delta(\alpha\beta\tilde{\phi}_{\beta,k} - n_p\beta\omega_k\tilde{\phi}_{\alpha,k} - \gamma\tilde{i}_{\beta,k}).
\end{aligned}
$$

5.3   NEURAL SLIDING MODES BLOCK CONTROL

Considering full-state measurements, the control objective is to track references for the mechanical rotor speed and square rotor flux magnitude of an induction motor, using the discrete-time neural model (5.1) and sliding mode control. Let us define the following vector:

$$
z_{1,k} = \bar{y}_k - y_{r,k}, \qquad (5.3)
$$

where $\bar{y}_k = [\hat{x}_{2,k}, \ \hat{x}_{3,k}]^\top$ and $y_{r,k} = [\omega_{r,k}, \ \Phi_{mr,k}]^\top$ are reference signals and are appropriate bounded signals with bounded increments. The dynamics of the new vector (5.3) is

$$
z_{1,k+1} = F_{z_1} = \begin{bmatrix} F_{z11} \\ F_{z12} \end{bmatrix}, \qquad (5.4)
$$

where

$$
\begin{aligned}
F_{z11} &= w_{21,k}S(w_k) - w'_{21}S(\phi_{\beta,k})i_{\alpha,k} + w'_{22}S(\phi_{\alpha,k})i_{\beta,k} - \omega_{r,k+1} \qquad (5.5) \\
F_{z12} &= w_{31,k}S(\Phi_{m,k}) + w'_{31}S(\phi_{\alpha,k})i_{\alpha,k} + w'_{32}S(\phi_{\beta,k})i_{\beta,k} - \Phi_{mr,k+1}.
\end{aligned}
$$

The relative degree for vector $z_{1,k}$ is two. Then, applying the block control technique and introducing a desired dynamics as

$$z_{1,k+1} = K_1 z_{1,k}, \qquad (5.6)$$

where $K_1 = diag\{k_{11}, k_{12}\}$ is a Schur matrix, then a new vector is defined as

$$z_{2,k} = F_{z_1} - K_1 z_{1,k}. \qquad (5.7)$$

Hence, the complete dynamics of z_k is as follows

$$
\begin{aligned}
z_{1,k+1} &= K_1 z_{1,k} + z_{2,k} \\
z_{2,k+1} &= F_{z_2} + B_{z_2} u_k,
\end{aligned}
\qquad (5.8)
$$

where $F_{z_2} = [F_{z_{21}}, \ F_{z_{22}}]^\top$, with

$$
\begin{aligned}
F_{z_{21}} =\ & w_{21,k+1} S(F_{z_{11}} + \omega_{r,k+1}) + w'_{21} S(\phi_{\beta,k+1})(w_{41,k} S(\phi_{\alpha,k}) \\
& + w_{42,k} S(\phi_{\beta,k}) + w_{43,k} S(i_{\alpha,k})) + w'_{22} S(\phi_{\alpha,k+1})(w_{51,k} S(\phi_{\alpha,k}) \\
& + w_{52,k} S(\phi_{\beta,k}) + w_{53,k} S(i_{\beta,k})) - \omega_{r,k+2} - k_{11} F_{z_{11}},
\end{aligned}
$$

$$
\begin{aligned}
F_{z_{22}} =\ & w_{31,k+1} S(F_{z_{12}} + \Phi_{mr,k+1}) + w'_{31} S(\phi_{\alpha,k+1})(w_{41,k} S(\phi_{\alpha,k}) \\
& + w_{42,k} S(\phi_{\beta,k}) + w_{43,k} S(i_{\alpha,k})) + w'_{32} S(\phi_{\beta,k+1})(w_{51,k} S(\phi_{\alpha,k}) \\
& + w_{52,k} S(\phi_{\beta,k}) + w_{53,k} S(i_{\beta,k})) - \Phi_{mr,k+2} - k_{12} F_{z_{12}},
\end{aligned}
$$

$$
B_{z_2} = \begin{bmatrix} w'_{21} w'_{41} S(\phi_{\beta,k+1}) & w'_{22} w'_{51} S(\phi_{\alpha,k}) \\ w'_{31} w'_{41} S(\phi_{\alpha,k+1}) & w'_{32} w'_{51} S(\phi_{\beta,k}) \end{bmatrix}.
$$

It is assumed that $S(\phi_{\alpha,k+1}) = S(\phi_{\alpha,k})$ and $S(\phi_{\beta,k+1}) = S(\phi_{\beta,k})$. For sliding mode control, the surface $s_k = 0$ is selected as $s_k = z_{2,k}$. In order to define the control law, a discrete-time sliding mode version [10], is implemented as:

$$
u_k = \begin{cases} u_{eq,k} & if \quad \|u_{eq}\| \leq u_0 \\ u_{o,k}\dfrac{u_{eq}}{\|u_{eq}\|} & if \quad \|u_{eq}\| > u_0 \end{cases}, \tag{5.9}
$$

where $u_{eq,k}$ is calculated from $s_{k+1} = 0$ of the form

$$
u_{eq,k} = -B_{z2}^{-1}[F_{z2}], \tag{5.10}
$$

and u_0 is bounded by the DC Link voltage in the inverter.

5.4 NEURAL INVERSE OPTIMAL CONTROL

Considering full-state measurements, the control objective is to track references for mechanical rotor speed and square rotor flux magnitude of an induction motor, using the discrete-time neural model (5.1) and inverse optimal control.

Hence, the inverse optimal control u_k

$$
u_k = \alpha(z_k) = -\left[I_m + J(z_k)\right]^{-1} h(z_k),
$$

with

$$
h(z_k) = g(z_k)^T P f(z_k) \tag{5.11}
$$

and

$$
J(z_k) = \frac{1}{2}g(z_k)^T P g(z_k) \tag{5.12}
$$

is applied to system (5.8), which steers $z_{2,k}$ to zero; where $f = F_{z_2}$, $g = B_{z_2}$, and $P \in \mathbb{R}^{2\times2}$ and $I \in \mathbb{R}^{2\times2}$ are positive definite and symmetric matrices. Therefore, the tracking error $z_{1,k}$ tends to zero as well and the cost functional (2.2) is minimized. Note that the control law u_k is bounded by the DC Link voltage as explained below.

5.5 IMPLEMENTATION

Rapid Control Prototyping (RCP) is a new and very important technology for speeding up product development time. The key element of RCP is automatic code generation, which eliminates error-prone hand coding procedures, thus making it possible for engineers to focus on control system synthesis, implementation, and evaluation, rather than on time-consuming low-level programming [10].

Different companies provide RCP software and hardware solutions. MATLAB and Simulink[1] are probably the best known and widely used simulation environments; furthermore, RCP enables control system design using block-diagram programming. Controller boards like DS1104 and DS1103[2] are appropriate for motion control and are fully programmable from the Simulink environment. At the present time there are different microcontrollers cheaper than the dSPACE boards, with similar characteristics.

TMS320F28069M[3] is a low-cost microcontroller for real-time control applications such as AC motor control, industrial drives, pumps, HVAC systems, solar inverters, digital power supplies, LED lighting, battery charging, smart grid and power line communications. It belongs to the C2000 MCU family with a 32-bit C28x DSP core and a real-time Control Accelerator (CLA) to increase the bandwidth of the C28x core, up to 240 MIPS of total performance [19]. TMS320F28069M is fully programmable from the Simulink environment.

DSP technology is widely applied to induction motors for different applications, e.g., methods of rotor fault detection in a closed-loop drive using neural networks [11], thermal monitoring methods using the estimated stator resistance as a direct indicator of average winding temperature [9, 7], comparison between PWM methods as space vector modulation (SVM) and nearest-level modulation [8], and comparison between control algorithms, e.g., SMC and field-oriented control (FOC) [4].

---

[1] MATLAB® and Simulink are registered trademarks of MathWorks Inc., Natick, Massachusetts, U.S.A.

[2] DS1104 and DS1103 are registered trademarks of dSPACE GmbH, Paderborn, Germany.

[3] TMS320F28069M is a registered trademark of Texas Instruments Inc., Dallas, Texas, U.S.A.

5.6   PROTOTYPE

In this section, a neural controller for a 5-HP induction motor is implemented on an RCP system, which is integrated with a TMS320F28069M microcontroller. The neural control scheme is implemented in real time using a prototype of the Intelligent Power Infrastructure Consortium laboratory at the Georgia Institute of Technology, with a TMS320F28069M microcontroller, in order to illustrate the effectiveness of the proposed neural controller and performance of the RCP system.

Figure 5.2 displays a picture of the prototype. The parameter values of the 5-HP induction motor are shown in Table 5.1. The DC Link voltage applied to the three-phase inverter comes from a three-phase rectifier and a three-phase transformer connected to the grid and it is regulated to 55 V, measured at the rectifier output. Figure 5.3 illustrates the internal configuration of the three-phase inverter.

TMS320F28069M and the devices connected to it, are illustrated in Figure 5.4. The instrumentation card is connected to the microcontroller to regulate the output voltages of the Hall effect current sensor. The optocoupler card is used to isolate the microcontroller from the power system. The FT232RL is used to convert serial transmissions to USB signals, in order to allow data transmission between the microcontroller and the computer.

5.6.1   RCP SYSTEM

TMS320F28069M is a fixed-point device with an IQMath virtual floating-point engine. It features high-resolution PWMs, 12-bit ADCs, Quadrature Encoder Pulse (QEP) interfaces, 12-bit DACs, GPIOs, and a host of serial communications such as dual CAN, SPI, SCI, and I2C. The implemented system diagram is illustrated in Fig. 5.1. The diagram illustrates all the parts of the control scheme such as the neural identifier, the neural controller and the flux observer; it includes also the Space Vector PWM (SVPWM), the ADC and the QEP driver.

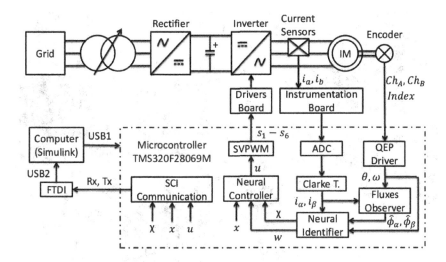

FIGURE 5.1   Implemented system diagram.

TABLE 5.1

Induction Motor Parameters

Parameter	Value	Description
P_{nom}	5 HP	Nominal power
V_{nom}	220 V	Nominal voltage
ω_{nom}	1745 rpm	Nominal speed
R_s	0.353 Ω	Stator resistance
R_r	0.424 Ω	Rotor resistance
L_m	33.7 mH	Mutual inductance
L_s	70.1 mH	Stator inductance
L_r	71.3 mH	Rotor inductance
n_p	2	Number of poles pairs
J	0.018 Kg m^2	Moment of inertia

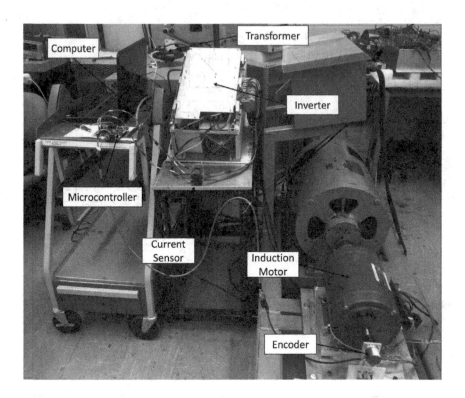

FIGURE 5.2   Prototype.

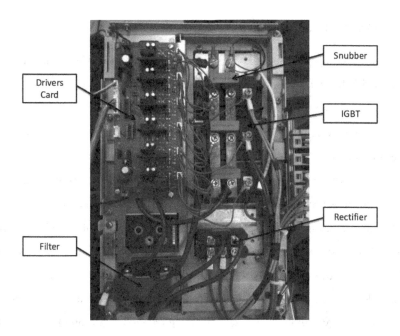

FIGURE 5.3   Inverter.

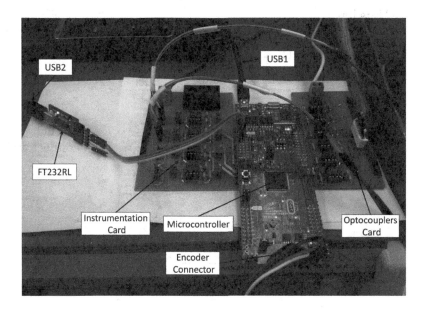

FIGURE 5.4   DSP and devices connected.

5.6.2 POWER ELECTRONICS

The electrical components of the prototype are connected according to the diagram
of Figure 5.1. It is worth noting that the DC Link voltage applied to the three-phase
inverter comes from a three-phase rectifier and a three-phase transformer connected
to the grid and it is regulated to 55 VDC, measured at the rectifier output. The three-
phase inverter circuit is illustrated in Figure 5.5. This inverter generates a sinusoidal
voltage signal with a variable frequency in order to control the states of an induction
motor as position, speed, torque and magnetic fluxes. The inverter is composed of six
IGBTs which have to be activated via PWM signals. The TMS320f28069M generates
the PWM signals; hence an optocoupler circuit is necessary to isolate the microcontroller from the power system; Figure 5.6 illustrates an isolated circuit using a 6N137
optocoupler. To activate the IGBT, a gate driver circuit is used (Figure 5.7); this circuit
is composed of a M57959L device, which operates as an isolation amplifier, provides
short circuit protection and a fault signal if the short circuit protection is activated.

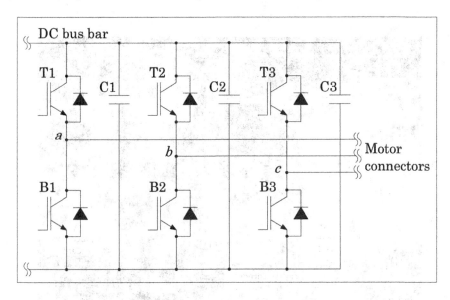

FIGURE 5.5   Three-phase inverter.

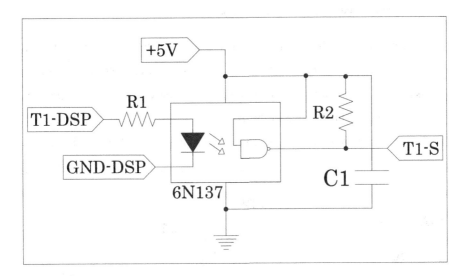

FIGURE 5.6   Optocoupler input driver circuit.

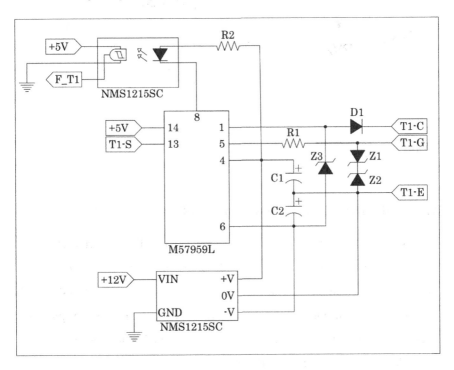

FIGURE 5.7   IGBT gate driver.

5.6.3 SIGNAL CONDITIONING FOR ADC

The phase currents of the induction motor are measured by Hall effect sensors (model: HAL 400-S). The voltage output of the sensor is $\pm 4V$ and the voltage range of the ADC is $[0\,V, 3.3\,V]$, then a signal conditioning circuit is necessary. Figure 5.8 presents a circuit to regulate the voltage in the ADC range using an operational amplifier OPA227P, which combines low noise and wide bandwidth with high precision.

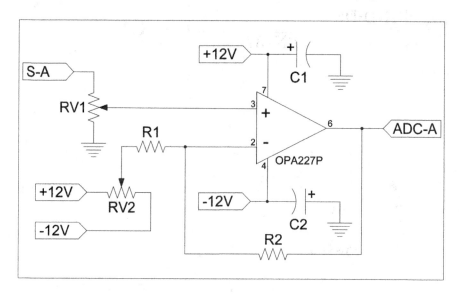

FIGURE 5.8   Signal conditioning circuit for phase currents.

5.6.4 REAL-TIME CONTROLLER IMPLEMENTATION

Simulink, Real-Time Workshop, and the embedded coder support package for Texas Instruments C2000 processors provide an integrated platform to design, simulate, implement, and verify embedded control systems on C2000 DSP boards as TMS320F28069M. Figure 5.6.4.1.2 illustrates the block diagram designed in a Simulink environment for the control scheme implementation of an induction motor. The block diagram is divided in 5 principal blocks:

· Induction motor inputs and outputs

- Fluxes observer
- Neural identifier
- Controller
- Serial communication interface

5.6.4.1 Induction Motor Inputs and Outputs

The first block contains input and output signals for the induction motor. The outputs of the induction motor block are rotor position, rotor speed and stator currents in components α, β, and the inputs are the voltages u_α, u_β.

5.6.4.1.1 Position and speed measurement via QEP

The position of the shaft is determined by the QEP, using the encoder signals: ch_A, ch_B and *Index*. The encoder has 1000 ppr and the resolution according to the possible combinations with two channels is calculated as

$$RES_{QEP} = \frac{2\pi}{(4 \cdot PPR) - 1} = 0.0016 \text{ rad.}$$

Rotor speed is calculated by the difference of the shaft position between each sampling time T_s.

5.6.4.1.2 Current measurement via ADC

The stator currents i_α, i_β are calculated from the phase currents i_a, i_b, i_c by the Clarke transform. Hence it is necessary to measure two currents (i_a, i_b), the third one is estimated as $i_c = -i_a - i_b$. The signal conditioning from the current sensor is in a range of $[0\ V, 3.3\ V]$ and there are 12 ADC bits; then, the respective resolution is given as

$$RES_{ADC} = \frac{V_{in}}{2^n - 1} = 80.58 \text{ mV.}$$

Then, an offset of 1.65 V is applied to the ADC signal in order to define a new origin and manipulate positive and negative voltages in the range of $[1.65\ V, -1.65\ V]$. Finally, this signal is multiplied by a gain to transform from voltage variables to current variables. Figure 5.10 illustrates the block diagram for current conditioning.

5.6.4.1.3 Voltage inputs determined via SVM

The control algorithm calculates voltage inputs for the induction motor as two sinusoidal signals u_α, u_β, which have to be converted into six PWM signals (T1, T2, T3, B1, B2 and B3), to activate the IGBTs of a three-phase inverter in order to generate three digital sinusoidal signals u_a, u_b, u_c. SVM is a technique to approximate the control voltage vector $u = \begin{bmatrix} u_\alpha & u_\beta \end{bmatrix}^\top$ instantaneously by combination of the switching states corresponding to the basic space vectors ($U_{0°}, U_{60°}, U_{120°}, U_{180°}, U_{240°}, U_{300°}$, O_0, O_1), which are illustrated in the space α, β plane in Figure 5.11 [22]. In order to estimate the average control voltage u in the function of the basic space vectors for any small sampling time T_s, the equation is defined as

$$\frac{1}{T_s} \int_{kT_s}^{(k+1)T_s} u(t) = \frac{1}{T_s} \left(\overline{T}_1 U_x + \overline{T}_2 U_{x\pm 60°} \right), \tag{5.13}$$

where $\overline{T}_1$ and $\overline{T}_2$ are the respective times to apply the switching states, corresponding to U_x and $U_{x\pm 60°}$ in the basic space vectors that form the sector containing u. If we assume that T_s is small enough, then Equation (5.13) becomes as

$$u_k = \frac{1}{T_s} \left(\overline{T}_1 U_x + \overline{T}_2 U_{x\pm 60°} \right). \tag{5.14}$$

Hence, $\overline{T}_1$ and $\overline{T}_2$ can be calculated as

$$\begin{bmatrix} \overline{T}_1 & \overline{T}_2 \end{bmatrix}^\top = T_s \begin{bmatrix} U_x & U_{x\pm 60°} \end{bmatrix}^{-1} u_k. \tag{5.15}$$

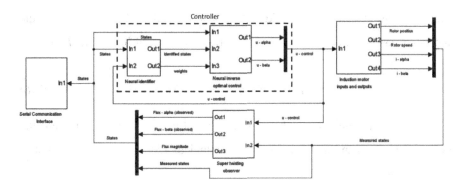

FIGURE 5.9   Block diagram for neural inverse optimal control implementation.

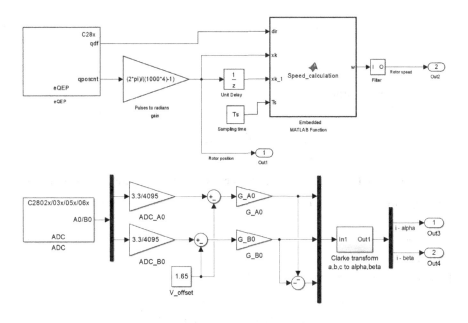

FIGURE 5.10   Block diagram for current conditioning.

Since the sum of $\overline{T}_1$ and $\overline{T}_2$ should be less than or equal to T_s ($\overline{T}_1 + \overline{T}_2 \leq T_s$), the inverter needs to be in O_0 or O_1 state for the rest of the period. Therefore, $\overline{T}_1 + \overline{T}_2 + \overline{T}_0 = T_s$ and

$$T_s u_k = \overline{T}_1 U_x + \overline{T}_2 U_{x \pm 60^\circ} + \overline{T}_0 (O_o \text{ or } O_1). \tag{5.16}$$

It is necessary to know in which sector the control voltage u_k is, to determine the switching time instants and sequence. In order to determine the sector of the control voltage vector, first let calculate

$$v_{ref1} = v_\beta$$
$$v_{ref2} = \sin(60^\circ)v_\alpha - \sin(30^\circ)v_\beta \tag{5.17}$$
$$v_{ref3} = -\sin(60^\circ)v_\alpha - \sin(30^\circ)v_\beta.$$

Secondly, let us determine

$$N = sign(v_{ref1}) + 2sign(v_{ref2}) + 4sign(v_{ref3}). \tag{5.18}$$

Thirdly, refer to Table 5.2 to map N to the sector of u_k. Figure 5.12 presents the waveform for each sector of a symmetric switching scheme.

TABLE 5.2
Determination of the Sector of u_k Based on N

N	1	2	3	4	5	6
Sector	II	VI	I	IV	III	V

It is only necessary to calculate the duty cycle for three IGBTs, the rest are complementaries as follows,

$$T_1 = 1 - B_1$$
$$T_2 = 1 - B_2$$
$$T_3 = 1 - B_3.$$

There are two forms to generate a PWM signal according to the compare function: symmetric and asymmetric. The PWM signals are configured with a symmetric compare function; this configuration is also known as continuous-up/down counting mode. The PWM compare function resolution can be computed with the period register value, which is estimated in the function of the PWM frequency 10 KHz and the CPU frequency 90 MHz as,

$$Period\_register = \frac{CPU\_frequency}{2 \cdot PWM\_frequency} = 4500.$$

Then, the approximate 12-bit resolution $2^{12} = 4096 \approx 4500$, the resolution in seconds is 0.122 ns and the PWM is configured with a dead time of 0.2 μs for short circuit protection. Figure 5.13 depicts the block diagram for space vector modulation.

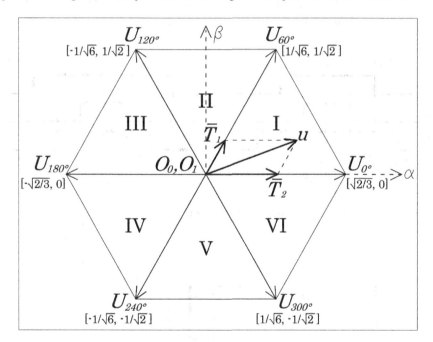

FIGURE 5.11   Control voltage vector in the space α,β plane.

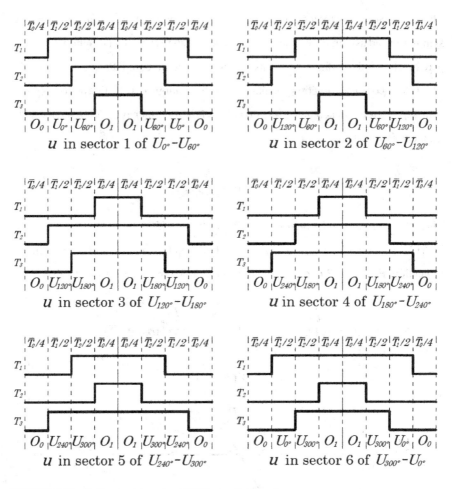

FIGURE 5.12   Software-determined SVM waveform pattern.

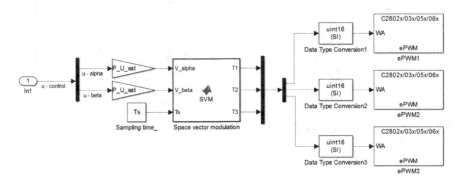

FIGURE 5.13   Block diagram for space vector modulation.

5.6.4.2 Flux Observer

The second principal block of the control block scheme (Figure 5.6.4.1.2) is a discrete-time super-twisting observer and is presented in Section 5.2. This observer estimates the magnetic flux components $\hat{\phi}_{\alpha,k}$, $\hat{\phi}_{\beta,k}$ and hence the square rotor flux magnitude $\Phi_{m,k} = \hat{\phi}_{\alpha,k}^2 + \hat{\phi}_{\beta,k}^2$. The estimated variables are in the function of the measured variables $\eta_k = \begin{bmatrix} \theta_k & \omega_k & i_{\alpha,k} & i_{\beta,k} \end{bmatrix}^\top$ and the control law u_k.

5.6.4.3 Neural Identifier

The neural identifier block estimates an accurate neural model of the induction motor in the function of the states x_k and the control law u_k, and is robust to external disturbances and parameter variations. The controllers are designed to track rotor speed and flux magnitude references in the function of the neural model. The neural identifier is described in Section 5.1.

5.6.4.4 Serial Communication Interface

The last principal block is the SCI, also known as the universal asynchronous receiver/transmitter (UART). The SCI transmits the state variables of the plant from the microcontroller to the computer. In order to allow data transmission, an FT232RL device is used to convert the serial signals (Rx and Tx) to USB signals. The configuration of the transmission is with a baud rate of 230400 bits/s, a free run mode and without parity mode, with the ability to transmit three 8-bit integer data.

5.6.5 NEURAL SLIDING MODE REAL-TIME RESULTS

These results are presented as follows. Figure 5.14 illustrates simulation identification results and Figure 5.15 displays the weights evolution. The sinusoidal voltage signal applied is $55\sin(40\pi t)$ and the training time is $15\ s$, which allows that the system reaches steady state. For tracking, the angular velocity reference is a sinusoidal signal $(17.5\sin(\pi t/15)+22.5)$ and $105\ s$ later is a square signal in the interval of $[20,40]$; the rotor flux reference is a constant signal $\Phi_{mr,k} = 0.1\ Wb^2$. The parameters used in the control law and observer are $\delta = 500 \, \mu s$, $K_1 = \begin{bmatrix} 0.8 & 0 \\ 0 & 0.5 \end{bmatrix}$, $c_1 = c_2 = c_3 = c_4 = 0.4$ and $c_5 = c_6 = c_7 = c_8 = 0.15$. Real-time tracking results for angular velocity and square rotor flux magnitude are presented in Figure 5.16 and Figure 5.17, respectively. Successful performance is obtained for both tracking outputs. It is worth noting that, flux magnitude tracking performance is adequate, which also means that the flux observer has good performance.

5.6.6   NEURAL INVERSE OPTIMAL CONTROL REAL-TIME RESULTS

In order to illustrate the performance of the NIOC, it is compared in real time with an IOC, which is a similar controller without the neural identifier. The experimental tests are performed for tracking rotor speed and magnetic flux references.

The respective real-time results are presented as follows. Figure 5.18 illustrates the rotor speed tracking performance and the tracking error for both controllers. It is significant that the tracking error is larger for the controller without the neural identifier (IOC). Figure 5.19 displays the controller comparison for the magnetic flux tracking performance and the tracking error. The performance for magnetic flux is similar for both controllers.

Table 5.3 presents the mean and the standard deviation for tracking errors. The rotor speed tracking error mean and rotor speed standard deviation for the IOC are larger than the NIOC. Therefore, the best rotor speed performance is obtained with the NIOC; this is due to the neural identifier, which estimates an accurate neural model. The magnetic flux tracking error mean is larger for the IOC than the NIOC too. Moreover, these tracking error means are small; therefore, the magnetic flux tracking performance is good in both controllers, which also means that the flux observer has a good performance.

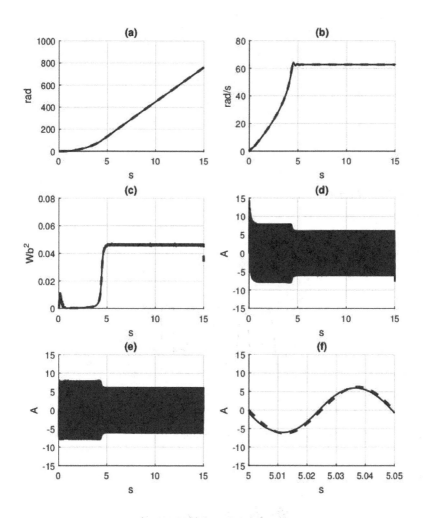

FIGURE 5.14 Identification results for the following states: (a) rotor angular position $\chi_{1,k}$; (b) rotor angular velocity $\chi_{2,k}$; (c) square rotor flux magnitude $\chi_{3,k}$; (d) stator current α $\chi_{4,k}$; (e) stator current β $\chi_{5,k}$; (f) zoom of (e) (neural signal in solid line and plant signal in dashed line).

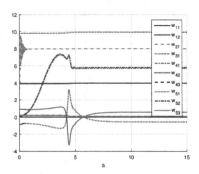

FIGURE 5.15   Weights evolution.

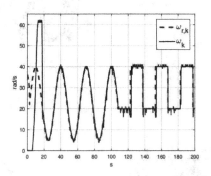

FIGURE 5.16   Angular velocity tracking performance.

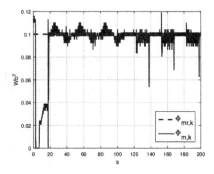

FIGURE 5.17   Square rotor flux magnitude tracking performance.

TABLE 5.3

Statistical Measures of Tracking Errors

Reference	Controller	Mean	Standard deviation
Rotor speed	NIOC	−0.3896	1.9422
Rotor speed	IOC	−0.5400	3.0369
Magnetic Flux	NIOC	−0.00053	0.0041
Magnetic Flux	IOC	−0.0011	0.0045

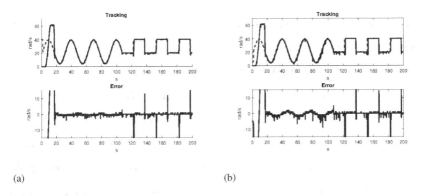

(a)   (b)

FIGURE 5.18   Comparison of rotor speed tracking performance and tracking error $\omega_k - \omega_{r,k}$ (output in solid line and reference in dashed line).

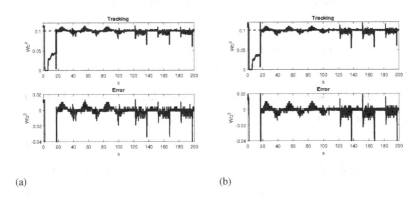

(a)   (b)

FIGURE 5.19   Comparison of square rotor flux magnitude tracking performance and tracking error $\Phi_{m,k} - \Phi_{mr,k}$ (output in solid line and reference in dashed line).

5.7 CONCLUSIONS

In this chapter, two control schemes are implemented in real time for a 5 HP induction motor. The schemes are implemented based on a neural identifier. The control schemes use a rapid control prototyping TMS320F28069M Microcontroller. Certainly, RCP meets the objectives of reducing development time and focusing on control system synthesis. Experimental results illustrate the effectiveness of the control schemes.

REFERENCES

1. A. Y. Alanis, E. N. Sanchez, and A. G. Loukianov. Discrete-time adaptive back-stepping nonlinear control via high-order neural networks. *Neural Networks, IEEE Transactions on*, 18(4):1185–1195, July 2007.

2. A. Y. Alanis, E. N. Sanchez, A. G. Loukianov, and M.A. Perez-Cisneros. Real-time discrete neural block control using sliding modes for electric induction motors. *Control Systems Technology, IEEE Transactions on*, 18(1):11–21, 2010.

3. M. E. Antonio-Toledo, E.N. Sanchez, and A.G. Loukianov. Real-time implementation of a neural block control using sliding modes for induction motors. In *World Automation Congress (WAC), 2014*, 1:502–507, Aug 2014.

4. A Benchaib, A Rachid, and E. Audrezet. Sliding mode input-output linearization and field orientation for real-time control of induction motors. *Power Electronics, IEEE Transactions on*, 14(1):3–13, Jan 1999.

5. F. Blaschke. A new method for the structural decoupling of a.c. induction machines. In *Conference Record IFAC*, Dubrovnik, Yugoslavia, 1971.

6. C.C. Chan. The state of the art of electric, hybrid, and fuel cell vehicles. *Proceedings of the IEEE*, 95(4):704–718, April 2007.

7. S. Cheng, Y. Du, J. A. Restrepo, P. Zhang, and T. G. Habetler. A nonintrusive thermal monitoring method for induction motors fed by closed-loop inverter drives. *Power Electronics, IEEE Transactions on*, 27(9):4122–4131, Sept 2012.

8. Y. Deng and R. G. Harley. Space-vector versus nearest-level pulse width modulation for multilevel converters. *Power Electronics, IEEE Transactions on*,

30(6):2962–2974, June 2015.

9. L. He, S. Cheng, Y. Du, R. G. Harley, and T. G. Habetler. Stator temperature estimation of direct-torque-controlled induction machines via active flux or torque injection. *Power Electronics, IEEE Transactions on*, 30(2):888–899, Feb 2015.

10. D. Hercog, M. Curkovic, and K. Jezernik. DSP-based rapid control prototyping systems for engineering education and research. In *Computer Aided Control System Design, IEEE Conference on*, 1:2292–2297, Munich, Germany, Oct 2006.

11. X. Huang, T. G. Habetler, and R. G. Harley. Detection of rotor eccentricity faults in a closed-loop drive-connected induction motor using an artificial neural network. *Power Electronics, IEEE Transactions on*, 22(4):1552–1559, July 2007.

12. A. G. Loukianov, J. Rivera, and J. M. CañÂ˜edo. Discrete-time sliding mode control of an induction motor. In *IFAC 15th Triennial World Congress*, 2002.

13. A. G. Loukianov J. Rivera B. Castillo-Toledo, S. Di Gennaro. Discrete time sliding mode control with application to induction motors. *Automatica*, 44:3036–3045, 2008.

14. R. Marino, S. Peresada, and P. Valigi. Adaptive input-output linearizing control of induction motors. *Automatic Control, IEEE Transactions on*, 38(2):208–221, Feb 1993.

15. E. Quintero-Manriquez, E. N. Sanchez, and R. A. Felix. Induction motor torque control via discrete-time sliding mode. In *2016 World Automation Congress (WAC)*, 1:1–5, July 2016.

16. I. Salgado, S. Kamal, I. Chairez, B. Bandyopadhyay, and L. Fridman. Supertwisting-like algorithm in discrete time nonlinear systems. In *The 2011 International Conference on Advanced Mechatronic Systems*, 1:497–502, Aug 2011.

17. E. N. Sanchez, A. Y. Alanis, and A. G. Loukianov. *Discrete-Time High Order Neural Control*. Springer Verlag, Berlin, Germany, 2008.

18. H. J. Shieh and K. Shyu. Nonlinear sliding-mode torque control with adaptive backstepping approach for induction motor drive. *Industrial Electronics, IEEE Transactions on*, 46(2):380–389, Apr 1999.

19. T I Support. C2000 real-time microcontrollers. Technical report, Texas Instru-

ments Incorporated, USA, 2015.

20. V. Utkin, J. Guldner, and M. Shijun. *Sliding Mode Control in Electro-mechanical Systems*. Automation and Control Engineering. Taylor & Francis, 1999.

21. Z. Yan, C. Jin, and V.I Utkin. Sensorless sliding-mode control of induction motors. *Industrial Electronics, IEEE Transactions on*, 47(6):1286–1297, Dec 2000.

22. Z. Yu. Space-vector PWM with TMS320C24x/f24x using hardware and software determined switching patterns. Technical report, Texas Instruments, 1999.

6 Doubly Fed Induction Generator

Effectiveness of the proposed neural control schemes are illustrated by means of the application to DFIG. This machine is one the most important generators used for the horizontal axis wind turbines, employed in wind power. The neural controller is constituted by two components: system identification and trajectory tracking, which are solved independently. The neuronal identifier is defined as a RHONN, trained with an EKF, used to identify the model DFIG and DC Link; after that, based on this neural model, the controllers are employed to achieve trajectory tracking even in the presence of undesired disturbances. Experimental results confirm applicability of the proposed control schemes.

Wind energy has many advantages, because it does not pollute and is an inexhaustible source of energy. The DFIG is a machine which has excellent performance and is commonly used for wind energy [1, 6]. The connection of the DFIG rotor with the electric grid, through a back-to-back converter, allows the rotor speed to vary while synchronizing the stator directly to a fixed frequency, which is achieved by controlling the rotor magnetic field by means of rotor currents supplied from a rotor side converter (RSC) [3]. The RSC is connected via a DC Link to a grid-side converter (GSC), which is, in turn, connected to the stator terminals directly or through a step-up transformer. Both RSC and GSC are four-quadrant converters which allow bidirectional flow of power; different techniques have been proposed to control the rotor currents for this configuration [7].

Different control algorithms have been proposed for this configuration. Various nonlinear approaches have been employed to control DFIG: sliding mode control

[2], feedback linearization [11], passivity-based control [5], so on. The previously mentioned control techniques for the DFIG are synthesized in the continuous-time framework. Other interesting control techniques for the DFIG in the DT framework are presented in [8, 9].

6.1 NEURAL IDENTIFIERS

On the basis of the models described in Appendix A, in this section, neural identifiers for the DFIG and DC Link are established. The extended Kalman filter (EKF) is used as the training algorithm for all the neural identifiers.

6.1.1 DFIG NEURAL IDENTIFIER

A neural identifier is used for DFIG modeling; it is assumed that all the state variables are measured. For identification, a RHONN is used as

$$\hat{\omega}_{r,k+1} = w_{11,k}S(\omega_{r,k})^2 + w_{12,k}S(\omega_{r,k}) + w_{13}\tau_{e,k}, \tag{6.1}$$

$$\hat{i}_{ds,k+1} = w_{21,k}S(i_{ds,k}) + w_{22,k}S(i_{qs,k})$$
$$+ w_{23,k}S(\omega_{r,k})S(i_{ds,k})S(i_{qs,k}) + w_{24}v_{dr,k}, \tag{6.2}$$

$$\hat{i}_{qs,k+1} = w_{31,k}S(i_{qs,k}) + w_{32,k}S(i_{ds,k})$$
$$+ w_{33,k}S(\omega_{r,k})S(i_{ds,k})S(i_{qs,k}) + w_{34}v_{qr,k}, \tag{6.3}$$

$$\hat{i}_{dr,k+1} = w_{41,k}S(i_{dr,k}) + w_{42,k}S(i_{qr,k})$$
$$+ w_{43,k}S(\omega_{r,k})S(i_{dr,k})S(i_{qr,k}) + w_{44}v_{dr,k}, \tag{6.4}$$

$$\hat{i}_{qr,k+1} = w_{51,k}S(i_{qr,k}) + w_{52,k}S(i_{dr})$$

$$+ w_{53,k}S(\omega_{r,k})S(i_{dr,k})S(i_{qr,k}) + w_{54}v_{qr,k}. \qquad (6.5)$$

The training is performed online, using a series-parallel configuration as displayed in Figure 6.1. All the neural network states are initialized as zero. The covariance matrices are initialized as diagonal, with nonzero elements as $P_1(0) = 920$, $R_1(0) = 900$, $Q_1(0) = 800$ for the rotor speed, and $P_i(0) = 800$, $R_i(0) = 800$, $Q_i(0) = 50$, $i = 2,3,4,5$ for the stator and rotor currents.

The choice of the parameters $P_i(0)$, $Q_i(0)$, $R_i(0)$ is an important step in the design of a neural controller, because it is based on the neural identifier.

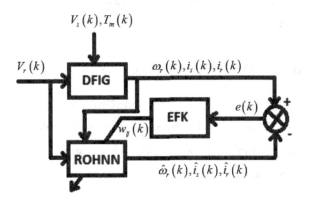

FIGURE 6.1   Identification scheme for the DFIG.

6.1.2   DC LINK NEURAL IDENTIFIER

A neural identifier is used for DC Link modeling; it is assumed that all the state variables are measured. For identification, a RHONN is defined as

$$\hat{v}_{dc,k+1} = w_{11}S(v_{dc,k}) + w_{12}S(v_{dc,k})S(i_{qg,k})$$

$$+ w_{13}i_{dg,k}, \qquad (6.6)$$

$$\hat{i}_{dg,k+1} = w_{21}S(i_{dg,k}) + w_{22}S(i_{qg,k}) + w_{23}S(v_{dc,k})$$

$$+ w_{24}v_{dg,k}, \tag{6.7}$$

$$\hat{i}_{qg,k+1} = w_{31}S(i_{qg,k}) + w_{32}S(i_{dg,k}) + w_{33}v_{qg,k}. \tag{6.8}$$

The training is performed online, using a series-parallel configuration. All the neural network states are initialized as zero. The covariance matrices are initialized as diagonal, with nonzero elements as $P_1(0) = 920$, $R_1(0) = 700$, $Q_1(0) = 100$ for the DC voltage, and $P_i(0) = 920$, $R_i(0) = 900$, $Q_i(0) = 100$, $i = 2,3$ for the currents.

Similar to the previous neural identifier, the choice of the initial conditions of $P_i(0)$, $Q_i(0)$, $R_i(0)$ is an important step in the design of a neural controller, because it is based on the neural identifier.

6.2 NEURAL SLIDING MODE BLOCK CONTROL

In this section, the procedure for design and development of controllers based on neural identifiers is detailed. The discrete sliding modes algorithm is used to develop the DFIG and the DC Link neural controllers, which are presented in Subsections 6.2.1 and 6.2.2, respectively.

6.2.1 DFIG CONTROLLER

The variables to be controlled are the DFIG electric torque ($\tau_{e,k}$) and the stator reactive power ($Q_{s,k}$). The control objectives are: a) to track a time-varying electric torque reference ($\tau_{e,k}^{ref}$), and b) to keep the electric power factor ($f_{ps1,k}$) at the stator terminals constant by means of the stator reactive power control. The control scheme utilized is shown in Figure 6.2. The electric torque ($\tau_{e,k}$) and stator reactive power ($Q_{s,k}$) are formulated, respectively, as

$$\tau_{e,k} = i_{r,k}^T M_{\tau_e} i_{s,k} \tag{6.9}$$

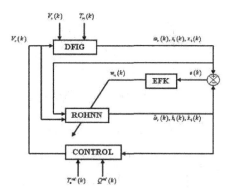

FIGURE 6.2 Neural control scheme.

and

$$Q_{s,k} = v_{s,k}^T M_Q i_{s,k}, \qquad (6.10)$$

where $M_{\tau_e} = X_m \begin{bmatrix} 0 & 1 \\ -1 & 0 \end{bmatrix}$, $M_Q = \begin{bmatrix} 0 & -1 \\ 1 & 0 \end{bmatrix}$.

The electric torque reference ($\tau_{e,k}^{ref}$) is defined as

$$\tau_{e,k}^{ref} = \gamma_{1,k}, \qquad (6.11)$$

where $\gamma_{1,k}$ is an arbitrary time-varying function, and the stator reactive power reference is defined as a power factor function ($f_{ps1,k}$):

$$Q_{s,k}^{ref} = \frac{P_{s,k}}{f_{ps1}} \sqrt{1 - f_{ps1}^2}, \qquad (6.12)$$

where $P_{s,k} \approx \tau_{e,k}^{ref}$.

In order to simplify the controller synthesis, the identifier equations are rewritten as

$$\hat{\omega}_{r,k+1} = \hat{f}_{\omega_{r,k}} + w_{13,k} \tau_{e,k}, \qquad (6.13)$$

$$\hat{x}_{1,k+1} = \hat{f}_{1,k} + B_1' u_k, \qquad (6.14)$$

$$\hat{x}_{2,k+1} = \hat{f}_{2,k} + B_2' u_k, \qquad (6.15)$$

where

$$\hat{x}_{1,k} = \begin{bmatrix} \hat{i}_{ds,k} \\ \hat{i}_{qs,k} \end{bmatrix}, \hat{x}_{2,k} = \begin{bmatrix} \hat{i}_{dr,k} \\ \hat{i}_{qr,k} \end{bmatrix},$$

$$\hat{f}_{\omega_{r,k}} = w_{11}S(\omega_{r,k})^2 + w_{12}S(\omega_k),$$

$$\hat{f}_{1,k} = \begin{bmatrix} w_{21}S(i_{ds,k}) + w_{22}S(i_{qs,k}) + \\ \dots w_{23}S(\omega_{r,k})S(i_{ds,k})S(i_{qs,k}) \\ w_{31}S(i_{qs,k}) + w_{32}S(i_{ds,k}) + \\ \dots w_{33}S(\omega_{r,k})S(i_{ds,k})S(i_{qs,k}) \end{bmatrix}$$

$$\hat{f}_{2,k} = \begin{bmatrix} w_{41}S(i_{dr,k}) + w_{42}S(i_{qr,k}) + \\ \dots w_{43}S(\omega_{r,k})S(i_{ds,k})S(i_{qs,k}) \\ w_{31}S(i_{qs,k}) + w_{32}S(i_{ds,k}) + \\ \dots w_{33}S(\omega_{r,k})S(i_{ds,k})S(i_{qs,k}) \end{bmatrix}$$

$$B_1' = \begin{bmatrix} w_{24} & 0 \\ 0 & w_{34} \end{bmatrix}, B_2' = \begin{bmatrix} w_{44} & 0 \\ 0 & w_{54} \end{bmatrix}.$$

In order to obtain the block system, from (6.9), (6.14), and (6.15), we can define

$$\hat{\tau}_{e,k+1} = \hat{f}_{\tau_e,k} + B_{\tau_e,k}u_k, \tag{6.16}$$

where

$$\hat{f}_{\tau_e,k} = \hat{f}_{2,k}^T M_{\tau_e} \hat{f}_{1,k} + u_k^T B_2'^T M_{\tau_e} B_1' u_k,$$

$$B_{T_e,k} = \hat{f}_{1,k}^T M_{\tau_e}^T B_2' + \hat{f}_{2,k}^T M_{\tau_e} B_1'.$$

For control synthesis, the input weights are assumed fixed and equal. Then, B_1' and B_2' are constant with equal entries; with these conditions, the nonlinear term $u_k^T B_2'^T M_{\tau_e} B_1' u_k = 0$ in $\hat{f}_{\tau_e,k}$, then $\hat{f}_{\tau_e,k} = \hat{f}_{2,k}^T M_{\tau_e} \hat{f}_{1,k}$. From (6.10), (6.14), and (6.15), we obtain the stator reactive power equation in differences as

$$\hat{Q}_{s,k+1} = \hat{f}_{Q_s,k} + B_{Q_s,k}u_k, \tag{6.17}$$

where

$$\hat{f}_{Q_s,k} = v_{s,k}^T M_{Q_s} \hat{f}_{1,k},$$

$$B_{Q_s,k} = v_{s,k}^T M_{Q_s} B'_{1,k}.$$

From (6.16) and (6.17), the block system is defined as

$$x_{1,k+1} = f_{x_1,k} + g_{x_1,k} u_k, \qquad (6.18)$$

where

$$x_{1,k} = \begin{bmatrix} \tau_{e,k} \\ Q_{s,k} \end{bmatrix}, \ f_{x_1,k} = \begin{bmatrix} \hat{f}_{\tau_e,k} \\ \hat{f}_{Q_s,k} \end{bmatrix},$$

$$g_{x_1,k} = \begin{bmatrix} B_{\tau_e,k} \\ B_{Q_s,k} \end{bmatrix}.$$

Clearly, (6.18) constitutes a one-block system [4]; then it can be used to synthetize the control input u_k as stated in the following theorem.

Theorem 6.1

For system (6.18), the control law defined as

$$u_k^c = u_k^{equ} + u_k^{din}$$

ensures trajectory tracking, with u_k^{equ} determined using discrete time sliding modes and u_k^{din}, a proportional term. ■

Proof

Using the discrete time sliding modes, the sliding manifold is defined as

$$s_k = x_{1,k} - x_{1,k}^{ref}, \tag{6.19}$$

where $x_{1,k}^{ref} = \begin{bmatrix} \tau_{e,k}^{ref} \\ Q_{s,k}^{ref} \end{bmatrix}.$

Evaluating (6.19) at $k+1$, the equivalent control u_k^{equ} is calculated as

$$u_k^{equ} = -g_{x_1,k}^{-1}(f_{x_1,k} - x_{1,k+1}^{ref}). \tag{6.20}$$

Applying $u_k = u_k^{equ}$ to (6.18), the state of the closed-loop system reaches the sliding manifold $s_k = 0$ in one sampling period. However, it is appropriate to add to the control signal a stabilizing term u_k^{equ}, in order to reach the sliding surface asymptotically and to avoid high gain control, defined as

$$u_k^{din} = -g_{x_1,k}^{-1}(Ks_k), \tag{6.21}$$

where K is a Schur matrix. Applying u_k in (6.18), the system dynamics of (6.19) are obtained as

$$s_{k+1} = Ks_k. \tag{6.22}$$

Then the sliding manifold $s_k = 0$ is reached asymptotically. ∎

To take into account the boundedness of the control signal $\|u_k\| < u_{max}$, $u_{max} > 0$, where $\|\cdot\|$ stands for the Euclidean norm, the following control law is selected:

$$u_k = \begin{cases} u_{max}\dfrac{u_k^c}{\|u_k^c\|} & if \ \|u_k^c\| > u_{max} \\ u_k^c & if \ \|u_k^c\| \le u_{max} \end{cases}. \tag{6.23}$$

6.2.1.1 Simulation Results

To evaluate the performance of the proposed controller, a simulation for a three-phase generator with a stator-referred rotor is developed. The generator parameters appear in Table 6.1.

The simulation conditions are:

- Simulation time: 12.5 seconds.
- Sampling time: $t_s = 0.5$ ms.
- DFIG initial conditions: rotor speed 0.3 pu, $i_{ds} = 0.001$ pu, $i_{qs} = 0.001$ pu, $i_{dr} = 0.001$ pu, $i_{qr} = 0.001$ pu.
- Identification input is a chirp signal, frequency range $0 - 60Hz$ and amplitude 0.1 pu.
- The 2.5-second initial signals are the identification; after that, the control signal is incepted at 2.5 seconds.
- The initial electric torque reference is a constant signal at 0.4 pu.
- From 1 to 3, seconds a pulse variation in the electric torque reference with amplitude of 0.5 pu is incepted.
- At 5 seconds, the electric torque reference is changed to a senoidal signal centered at 0.5 pu with amplitude of 0.4 pu and 1 Hz.
- Power factor reference is constant in 0.9.
- The gain K in (6.21) is defined as $\begin{bmatrix} 0.95 & 0 \\ 0 & 0 \end{bmatrix}$.

The behavior of the neural identifier is shown in Figure 6.3 to Figure 6.7. In these figures, the DFIG variables are presented jointly with their identifiers. It can be seen that all the identification errors are small; additionally, all the neural network weights are bounded as shown in part (b) of Figure 6.3 to Figure 6.7. At 2.5 seconds the control signals are incepted. Figure 6.8(a) presents the electric torque τ_e. In this figure, it can be seen that the tracking for electric torque is reached quickly. In Figure 6.8(b), the reactive power is presented. The references tracking are ensured by the neural controller, even if the references are time-varying signals. In Figure 6.8(c), the

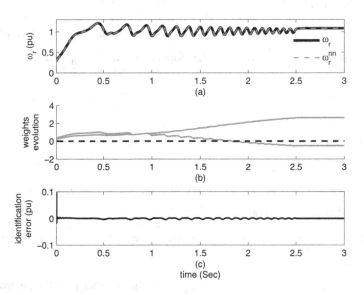

FIGURE 6.3　(a) Rotor speed (ω_r) with rotor speed identifier, (b) neural network weights evolution, and (c) identification error, respectively.

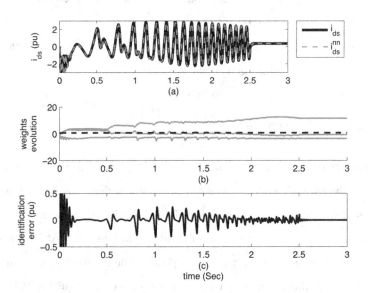

FIGURE 6.4　(a) Stator current (i_{ds}) with stator current identifier, (b) neural network weights evolution, and (c) identification error, respectively.

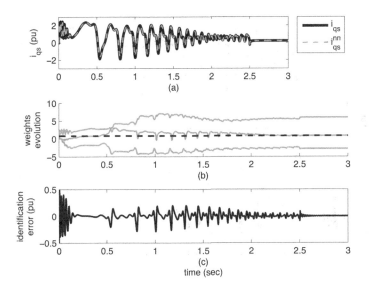

FIGURE 6.5   (a) Stator current (i_{qs}) with stator current identifier, (b) neural network weights evolution, and (c) identification error, respectively.

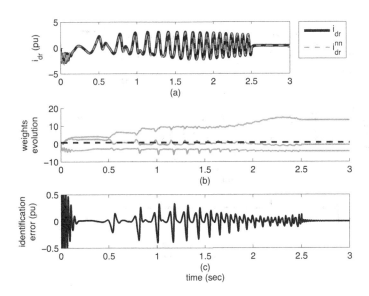

FIGURE 6.6   (a) Rotor current (i_{dr}) with rotor current identifier, (b) neural network weights evolution, and (c) identification error, respectively.

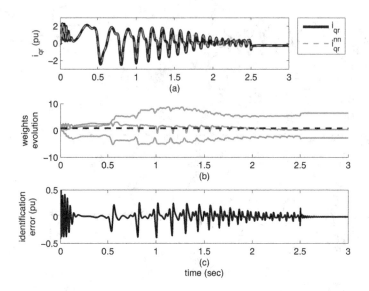

FIGURE 6.7   (a) Rotor current (i_{qr}) with rotor current identifier, (b) neural network weights evolution, and (c) identification error, respectively.

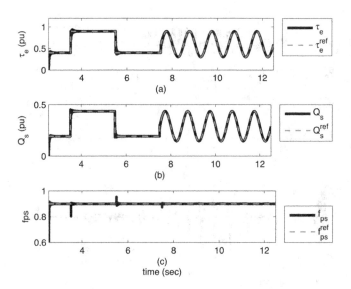

FIGURE 6.8   System outputs: (a) electric torque (τ_e) tracking, (b) reactive power (Q_s) tracking, and (c) power factor (f_{ps1}) tracking.

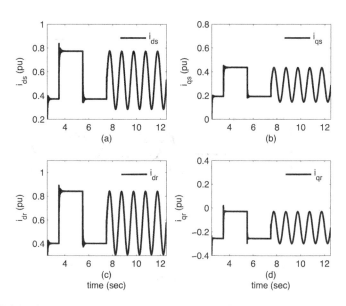

FIGURE 6.9 Generator currents: (a) stator current i_{ds}, (b) stator current i_{qs}, (c) rotor current i_{dr}, and (d) rotor current i_{qr}.

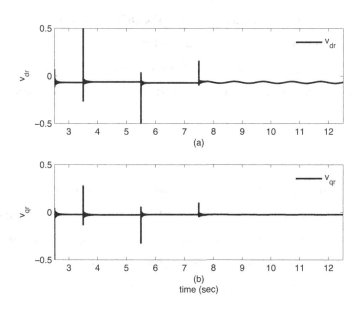

FIGURE 6.10 Control signals: (a) v_{dr} and (b) v_{qr}.

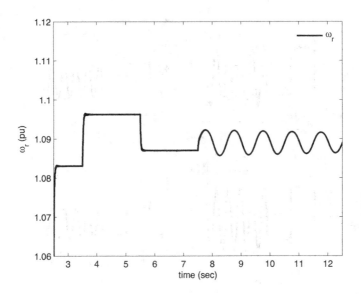

FIGURE 6.11   Rotor speed (ω_r).

power factor is displayed, and it can be seen that the tracking is good enough in the presence of electric torque reference variations. The DFIG current performances are shown in Figure 6.9, where we can see that the DFIG currents are within nominal limits. The control signals are bounded as shown in Figure 6.10. In Figure 6.11, the rotor speed is presented, which has small variations due to the electric torque tracking error.

6.2.2   DC LINK CONTROLLER

The variables to be controlled are the capacitor voltage ($v_{dc,k}$) and the reactive power ($Q_{g,k}$). The control objectives are: a) to track a DC voltage reference ($v_{dc,k}^{ref}$) on the DC Link, and b) to keep constant the electric power factor ($f_{ps2,k}$) at the step-up transformer terminals by means of the reactive power ($Q_{g,k}$) control.

The step-up transformer reactive power ($Q_{g,k}$) is formulated as

$$Q_{g,k} = v_{sg,k}^T M_Q i_{g,k}.$$
(6.24)

The DC voltage reference ($v_{dc,k}^{ref}$) is defined as

$$v_{dc,k}^{ref} = \gamma_{2,k}, \tag{6.25}$$

where $\gamma_{2,k}$ is an arbitrary time-varying function, and the reference for the reactive power is defined as a function of the electric power factor (f_{ps2}):

$$Q_{g,k}^{ref} = \frac{P_{g,k}}{f_{ps2}} \sqrt{1 - f_{ps2}^2}, \tag{6.26}$$

where $P_{g,k} = v_{sg,k}^T M_P i_{g,k}$, and $M_P = \begin{bmatrix} 1 & 0 \\ 0 & 1 \end{bmatrix}$.

Let us define the tracking error for the DC voltage as

$$\varepsilon_{1,k}^g = v_{dc,k} - v_{dc,k}^{ref}. \tag{6.27}$$

From (6.27), using (6.6), then $\varepsilon_{1,k+1}^g$ is equal to

$$\varepsilon_{1,k+1}^g = w_{11} S(v_{dc,k}) + w_{12} S(v_{dc,k}) S(i_{qg,k}) + w_{13} i_{dg,k} - v_{dc,k+1}^{ref}, \tag{6.28}$$

where it can be seen that the DC voltage ($v_{dc,k}$) is controlled directly by $i_{dg,k}$.

Then, the i_{dg} reference is defined as

$$i_{dg,k}^{ref} = w_{13}^{-1} \left(v_{dc,k+1}^{ref} + k_1 \varepsilon_{1,k}^g - w_{11} S(v_{dc,k}) - w_{12} S(v_{dc,k}) S(i_{qg,k}) \right), \tag{6.29}$$

where $k_1 \varepsilon_{1,k}^g$ is introduced to reach the reference asymptotically, with $|k_1| < 1$. On the other hand, the tracking error for the reactive power is

$$\varepsilon_{2,k}^g = Q_{g,k} - Q_{g,k}^{ref}. \tag{6.30}$$

From (6.24), and considering that $v_{qgs,k} = 0$, it could be established that

$$Q_{g,k} = -v_{dgs,k} i_{qg,k}. \tag{6.31}$$

In order to determine the reference (i_{qg}^{ref}), we assume that $\varepsilon_{2,k}^g = 0$, then $Q_{g,k} = Q_{g,k}^{ref}$; therefore, it is easy to see that i_{qg}^{ref} is given by

$$i_{qg,k}^{ref} = -i_{dg,k}^{ref}\frac{\sqrt{1-f_{ps2}^2}}{f_{ps2}}. \tag{6.32}$$

Considering $x_{2,k} = \begin{bmatrix} \hat{i}_{dg,k} \\ \hat{i}_{qg,k} \end{bmatrix}$, Equations (6.7) and (6.8) can be rewritten as follows:

$$x_{2,k+1} = f_{x_2,k} + g_{x_2}u_{g,k} \tag{6.33}$$

with

$$f_{x_2,k} = \begin{bmatrix} w_{21}S(i_{dg,k}) + w_{22}S(i_{qg,k}) + w_{23}S(v_{dc,k}) \\ w_{31}S(i_{qg,k}) + w_{32}S(i_{dg,k}) \end{bmatrix},$$

$$g_{x_2} = \begin{bmatrix} w_{24} & 0 \\ 0 & w_{33} \end{bmatrix}.$$

Then, it is clear that Equation (6.33) is of the form (6.18); according to Theorem 6.1, the control input $u_{g,k}$ is selected as follows.

At first, the sliding manifold is formulated as

$$s_{g,k} = x_{2,k} - x_{2,k}^{ref}, \tag{6.34}$$

where $x_{2,k}^{ref} = \begin{bmatrix} i_{dg,k}^{ref} \\ i_{qg,k}^{ref} \end{bmatrix}$.

Evaluating (6.34) at $(k+1)$ and using (6.33),

$$s_{g,k+1} = f_{x_2,k} + g_{x_2}u_{g,k} - x_{2,k+1}^{ref}. \tag{6.35}$$

Then, the equivalent control $u_{g,k}^{equ}$ is calculated as

$$u_{g,k}^{equ} = -g_{x_2}^{-1}(f_{x_2,k} - x_{2,k+1}^{ref}). \tag{6.36}$$

Applying $u_k = u_{g,k}^{equ}$ to the system, the state of the closed-loop system reaches the sliding manifold $s_{g,k} = 0$ in one sample time. However, it is appropriate to add to the control signal a stabilizing term $u_{g,k}^{din}$ in order to reach the sliding surface asymptotically and to avoid high gain control; hence, the complete control $u_{g,k}^c$ is proposed as

$$u_{g,k}^c = u_{g,k}^{equ} + u_{g,k}^{din},$$ (6.37)

where

$$u_{g,k}^{din} = g_{x_2}^{-1}\left(K_g s_{g,k}\right)$$ (6.38)

and $K_g = \begin{bmatrix} k_1^g & 0 \\ 0 & k_2^g \end{bmatrix}$ is a Schur matrix. To take into account the boundedness of the control signal $\left\|u_{g,k}\right\| < u_{g\,max}$, $u_{g\,max} > 0$, where $\|\cdot\|$ stands for the Euclidean norm, the following control law is selected:

$$u_{g,k} = \begin{cases} u_{g\,max}\dfrac{u_{g,k}^c}{\left\|u_{g,k}^c\right\|} & if \ \left\|u_{g,k}^c\right\| > u_{g\,max} \\[2ex] u_{g,k}^c & if \ \left\|u_{g,k}^c\right\| \leq u_{g\,max} \end{cases}.$$ (6.39)

The stability proof using (6.39) is presented in [10].

6.2.2.1 Simulation Results

To evaluate the performance of the proposed controller, a simulation for a DC Link is developed. The DC Link parameters appear in Table 6.1. The simulation is performed in MATLAB/Simulink, with conditions:

- Simulation time: 10 seconds.
- Sampling time: $t_s = 0.5 \ ms$.
- DC Link initial conditions: $v_{dc} = 0.01 \ pu$, $i_{dg} = 0 \ pu$, $i_{qg} = 0 \ pu$.
- Identification input is a chirp signal, frequency range 0–60Hz, and amplitude 0.01 pu.
- The first 1 second is the identification; after that, the control signal is incepted

at 1 second.

- The DC voltage reference is a constant signal at 0.5567 *pu*.
- Power factor reference is constant at 0.9 *pu*.
- Load resistance R_L of 383.0579 *pu* is connected to the capacitor in parallel scheme in order to simulate an unknown perturbation.

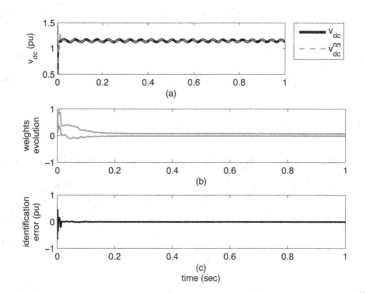

FIGURE 6.12   (a) DC voltage (v_{dc}) with the respective identifier, (b) neural network weights evolution, and (c) identification error.

The behavior of the neural identifier is shown in Figure 6.12 to Figure 6.14. In these figures, the DC Link variables are presented jointly with their identifiers; it can be seen that all the identification errors are small; additionally, all the neural network weights are bounded as shown in part (b) of Figure 6.12 to Figure 6.14. At 1 second the control signal is incepted.

The performance of the DC Link controller is shown in Figure 6.15 to Figure 6.17. The DC voltage (Figure 6.15(a)) and power factor tracking (Figure 6.15(b)) is controlled to the reference. In this figure, we can see that the DC voltage reaches the reference quickly, and the electric power factor is kept constant at 0.9 during the 10-second lapse. The control signals v_{dg} and v_{qg} are bounded as shown in Figure 6.16.

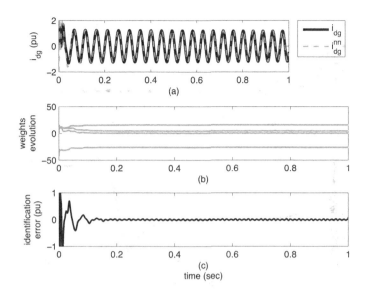

FIGURE 6.13 (a) DC Link current (i_{dg}) with the respective identifier, (b) neural network weights evolution, and (c) identification error.

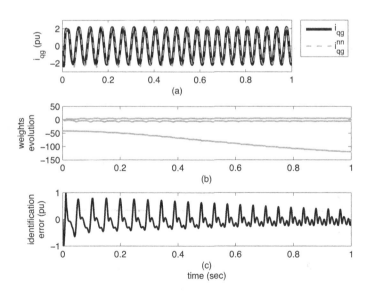

FIGURE 6.14 (a) DC Link current (i_{qg}) with the respective identifier, (b) neural network weights evolution, and (c) identification error.

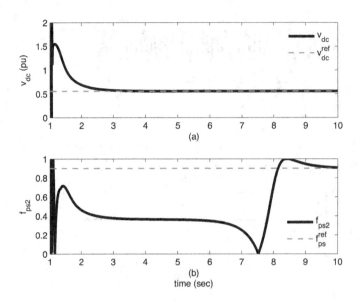

FIGURE 6.15   System outputs: (a) DC voltage (v_{dc}) and (b) step-up transformer power factor (f_{ps2}).

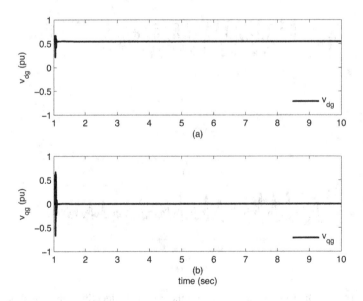

FIGURE 6.16   Control signals: (a) v_{dg} and (b) v_{qg}.

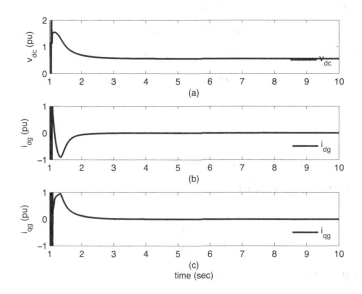

FIGURE 6.17   State variables: (a) DC voltage v_{dc}, (b) current i_{dg}, and (c) current i_{qg}.

The state variables of the DC Link are shown in Figure 6.17, where it can be seen that the state variables are stable and bounded. The transient lapse is short and the resistance R_L is connected in a parallel scheme to simulate the capacitor discharge by a resistive load.

6.3   NEURAL INVERSE OPTIMAL CONTROL

In this section, the inverse optimal control is based on the neural identifiers for the DFIG and DC Link controller development.

6.3.1   DFIG CONTROLLER

The variables to be controlled are the DFIG electric torque ($\tau_{e,k}$) and the stator reactive power ($Q_{s,k}$). The control objectives are: a) to track an electric torque trajectory ($\tau_{e,k}^{ref}$), and b) to keep the electric power factor ($f_{ps1,k}$) at the stator terminals constant by means of the stator reactive power control. The electric torque ($\tau_{e,k}$) and stator reactive power ($Q_{s,k}$) are defined, respectively, as (6.9) and (6.10).

In order to apply the inverse optimal control, we need to calculate the reference for the system state, which is obtained by considering the system steady state as the desired references, and then achieving trajectory tracking as a stabilization problem; hence the system steady state is obtained as follows. The electric torque reference $(\tau_{e,k}^{ref})$ is defined as

$$\tau_{e,k}^{ref} = \gamma_{1,k}, \tag{6.40}$$

where $\gamma_{1,k}$ is an arbitrary time-varying function, and the reactive power reference is defined as a function of electric power factor (f_{ps1}):

$$Q_{s,k}^{ref} = \frac{P_{s,k}}{f_{ps1}}\sqrt{1 - f_{ps1}^2}, \quad P_{s,k} \approx \tau_{e,k}^{ref}. \tag{6.41}$$

The tracking errors are defined, respectively, as

$$\varepsilon_{\tau_e,k} = \tau_{e,k} - \tau_{e,k}^{ref}, \tag{6.42}$$

$$\varepsilon_{Q_s,k} = Q_{s,k} - Q_{s,k}^{ref}. \tag{6.43}$$

In order to calculate the steady state, it is assumed that $\varepsilon_{\tau_e,k} = 0$ and $\varepsilon_{Q_s,k} = 0$; then using (6.9) and (6.10), equations (6.42) and (6.43) can be rewritten as

$$i_{r,k}^{ss\,T} M_{\tau_e} i_{s,k}^{ss} = \tau_{e,k}^{ref}, \tag{6.44}$$

$$v_{s,k}^{T} M_{Q_s} i_{s,k}^{ss} = Q_{s,k}^{ref}, \tag{6.45}$$

where the superscript ss denotes steady state. From (6.46) and (6.47),

$$i_{s,k+1} = i_{s,k} + t_s\left(A_{11,k}i_{s,k} + A_{12,k}i_{r,k}\right) + t_s(D_1 v_{s,k} + B_{1u,k}), \tag{6.46}$$

$$i_{r,k+1} = i_{r,k} + t_s(A_{21,k}i_{s,k} + A_{22,k}i_{r,k}) + t_s(D_2 v_{s,k} + B_{2u,k}), \tag{6.47}$$

the relation between stator and rotor currents in steady state is obtained as

$$i_{r,k}^{ss} = G_1 i_{s,k}^{ss} + H_1 v_{s,k}, \tag{6.48}$$

$$i_{s,k}^{ss} = G_2 i_{r,k}^{ss} + H_2 v_{s,k}, \tag{6.49}$$

where G_1, H_1, G_2, and H_2 are defined. Solving i_{ds}^{ss}, i_{qs}^{ss}, i_{dr}^{ss}, and i_{qr}^{ss} of (6.44), (6.45), (6.48), and (6.49), the i_k^{ss} is defined as

$$i_k^{ss} = \begin{bmatrix} i_{ds,k}^{ss} & i_{qs,k}^{ss} & i_{dr,k}^{ss} & i_{qr,k}^{ss} \end{bmatrix}^T. \tag{6.50}$$

Now, based on the DFIG neural identifier, the inverse optimal controller is developed. In order to simplify the controller synthesis, the identifier equations can be rewritten as:

$$\hat{\omega}_{r,k+1} = f_{\omega r,k} + w_{13,k})T_{e,k}, \tag{6.51}$$

$$\hat{i}_{s,k+1} = f_{1,k} + B'_{1,k} u_k, \tag{6.52}$$

$$\hat{i}_{r,k+1} = f_{2,k} + B'_{2,k} u_k, \tag{6.53}$$

where

$$\hat{i}_{s,k} = \begin{bmatrix} \hat{i}_{ds,k} \\ \hat{i}_{qs,k} \end{bmatrix}, \hat{i}_{r,k} = \begin{bmatrix} \hat{i}_{dr,k} \\ \hat{i}_{qr,k} \end{bmatrix},$$

$$f_{\omega r,k} = w_{11} S(\omega_{r,k})^2 + w_{12} S(\omega_k),$$

$$f_{1,k} = \begin{bmatrix} w_{21} S(i_{ds,k}) + w_{22} S(i_{qs,k}) + \\ \dots w_{23} S(\omega_{r,k}) S(i_{ds,k}) S(i_{qs,k}) \\ w_{31} S(i_{qs,k}) + w_{32} S(i_{ds,k}) + \\ \dots w_{33} S(\omega_{r,k}) S(i_{ds,k}) S(i_{qs,k}) \end{bmatrix},$$

$$f_2(k) = \begin{bmatrix} w_{41}S(i_{dr,k}) + w_{42}S(i_{qr,k}) + \\ \dots w_{43}S(\omega_{r,k})S(i_{ds,k})S(i_{qs,k}) \\ w_{31}S(i_{qs,k}) + w_{32}S(i_{ds,k}) + \\ \dots w_{33}S(\omega_{r,k})S(i_{ds,k})S(i_{qs,k}) \end{bmatrix},$$

$$B'_{1,k} = \begin{bmatrix} w_{24,k} & 0 \\ 0 & w_{34,k} \end{bmatrix},$$

$$B'_{2,k} = \begin{bmatrix} w_{44,k} & 0 \\ 0 & w_{54,k} \end{bmatrix}.$$

The system denoted by (6.52)) and (6.53) is rewritten as

$$\hat{i}_{k+1} = \hat{f}_{rsc,k} + \hat{g}_{rsc,k} u_k, \tag{6.54}$$

where

$$\hat{f}_{rsc,k} = \begin{bmatrix} f_{1,k} \\ f_{2,k} \end{bmatrix}, \quad \hat{g}_{rsc,k} = \begin{bmatrix} B'_{1,k} \\ B'_{2,k} \end{bmatrix}.$$

In order to apply inverse optimal control, the tracking error is defined as

$$i_{e,k} = \hat{i}_k - i_k^{ss}. \tag{6.55}$$

Evaluating (6.55) at time $k+1$ and using (6.54), the error dynamic is given by

$$i_{e,k+1} = \hat{i}_{k+1} - i_{k+1}^{ss}$$

$$i_{e,k+1} = \hat{f}_{rsc,k} + \hat{g}_{rsc,k} u_k - i_{k+1}^{ss}. \tag{6.56}$$

For system (6.56), the control signal (u_k) is decomposed into two components:

$$u_k = u_{1,k} + u_{i,k}^*, \tag{6.57}$$

and in order to convert (6.56) into the form (2.1), $u_{1,k}$ is selected as

$$u_{1,k} = \hat{g}_{rsc,k^*-1}\left(i_{k+1}^{ss}\right),$$

where $\hat{g}_{rsc,k^*-1} = \left(\hat{g}_{rsc,k^T}\,\hat{g}_{rsc,k}\right)^{-1}\hat{g}_{rsc,k^T}$. Then, system (6.56), with (6.57) as input, results in

$$i_{e,k+1} = \hat{f}_{rsc,k} + \hat{g}_{rsc,k}\,u_{i,k}^*, \tag{6.58}$$

which has the form of (2.1) as

$$x_{k+1} = f_i(x_k) + g_i(x_k)\,u_{i,k}^*,$$

where $x_k := i_{e,k}$ is the system state, $f_i(x_k) = \hat{f}_{rsc,k}$, $g_i(x_k) = \hat{g}_{rsc,k}$, and hence the inverse optimal control law $(u_{i,k}^*)$ is established using Theorem 4.1 as

$$
\begin{aligned}
u_{i,k}^* = &-\frac{1}{2}\left(R_1(x_k) + \frac{1}{2}g_i^T(x_k)P_1\right.\\
&\left.\times g_i(x_k)\right)^{-1}g_i^T(x_k)P_1\,f_i(x_k),
\end{aligned}\tag{6.59}
$$

where for the rotor side converter (RSC) controller, $R_1(x_k)$ and P_1 are the $R(x_k)$ and P matrices, respectively.

6.3.1.1 Simulation Results

To evaluate the performance of the proposed controller, a simulation for a three-phase generator with a stator-referred rotor is developed. The generator parameters appear in Table 6.1.

The simulation conditions are:

- Simulation time: 12.5 seconds.
- Sampling time: $t_s = 0.5\ ms$.
- DFIG initial conditions: rotor speed 0.3 pu, $i_{ds} = 0.001\ pu$, $i_{qs} = 0.001\ pu$, $i_{dr} = 0.001\ pu$, $i_{qr} = 0.001\ pu$.

- Identification input is a chirp signal, frequency range 0–60 Hz, and amplitude 0.1 pu.
- The 2.5-second initial signal is the identification; after that, the control signal is incepted at 2.5 seconds.
- The initial electric torque reference is a constant signal at 0.4 pu.
- From 1 to 3 seconds, a pulse variation in the electric torque reference with amplitude of 0.5 pu is incepted.
- At 5 seconds the electric torque reference is changed to a senoidal signal centered at 0.5 pu with amplitude of 0.4 pu and 1 Hz.
- Power factor reference is constant at 0.9.

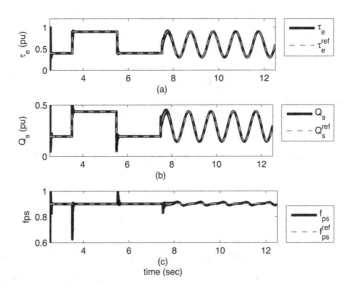

FIGURE 6.18   System outputs: (a) electric torque (τ_e) tracking, (b) reactive power (Q_s) tracking, and (c) power factor (f_{ps1}) tracking.

The behavior of the neural identifier is shown in Figure 6.3 to Figure 6.7. In the first 2.5 seconds the identification is achieved; after that, the control signals are incepted. Figure 6.18 presents the electric torque (τ_e) tracking (a), the reactive power (Q_s) tracking (b), and electric power factor (c). In this figure, it can be seen that the tracking for electric torque and reactive power are reached fast. The reference trackings are

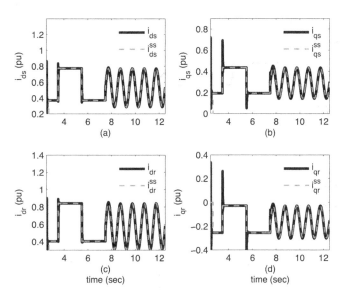

FIGURE 6.19   Generator currents: (a) stator current i_{ds}, (b) stator current i_{qs}, (c) rotor current i_{dr}, and (d) rotor current i_{qr}.

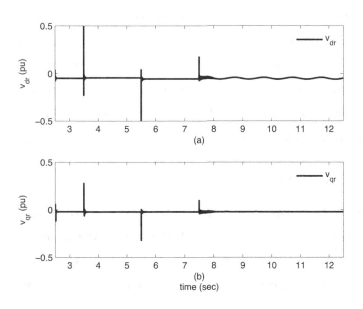

FIGURE 6.20   Control signals: (a) v_{dr} and (b) v_{qr}.

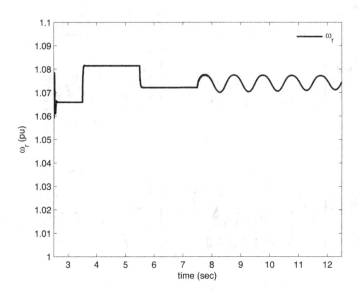

FIGURE 6.21   Rotor speed (ω_r).

ensured by the neural control algorithm despite the fact that the references are time-varying signals. The power factor is kept constant when the electric torque reference is a pulse signal; and the power factor has a small variation when the electric torque reference is a senoidal signal. The DFIG current performances are shown in Figure 6.19, where we can see that the DFIG currents reach their respective references. The control signals are bounded and these are shown in Figure 6.20. Rotor speeds associated with this experiment are shown in Figure 6.21.

6.3.2   DC LINK CONTROLLER

The variables to be controlled are the capacitor voltage ($v_{dc,k}$) and the reactive power ($Q_{g,k}$). The control objectives are: a) to track a DC voltage reference ($v_{dc,k}^{ref}$) for the DC Link, and b) to keep the electric power factor ($f_{ps2,k}$) constant at the step-up transformer terminals by means of the reactive power control ($Q_{g,k}$). The step-up transformer reactive power ($Q_{g,k}$) is defined as

$$Q_{g,k} = v_{sg,k}^T M_Q i_{g,k}.$$ (6.60)

The DC voltage reference ($v_{dc,k}^{ref}$) is defined as

$$v_{dc,k}^{ref} = \gamma_{2,k},$$ (6.61)

where $\gamma_{2,k}$ is an arbitrary time-varying function, and the reference for the reactive power is defined as a function of electric power factor (f_{ps2}):

$$Q_{g,k}^{ref} = \frac{P_{g,k}}{f_{ps2}} \sqrt{1 - f_{ps2}^2}, \ P_{g,k} = v_{sg,k}^T M_P i_{g,k}.$$ (6.62)

Similar to the previous case (DFIG controller), it is required to determine the steady-state values for v_{dc}^{ref} and Q_g^{ref}. The tracking error for the DC voltage and the reactive power are defined, respectively, as

$$\varepsilon_{v_{dc},k} = v_{dc,k} - v_{dc,k}^{ref}$$ (6.63)

$$\varepsilon_{Q_g,k} = Q_{g,k} - Q_{g,k}^{ref}.$$ (6.64)

Evaluating (6.63) at time $k+1$ and using (6.65),

$$v_{dc,k+1} = v_{dc,k} + t_s \left(\frac{1}{Cv_{dc,k}} v_{gs,k}^T M_P i_{g,k} \right)$$ (6.65)

yields

$$\varepsilon_{v_{dc},k+1} = v_{dc,k} + \frac{t_s}{Cv_{dc,k}} v_{gs,k}^T M_P i_{g,k} - v_{dc,k+1}^{ref}.$$ (6.66)

Assuming that in steady state $\varepsilon_{v_{dc},k} = \varepsilon_{v_{dc},k+1} = \varepsilon_{Q_g,k} = \varepsilon_{Q_g,k+1} = 0$, then $v_{dc,k}^{ss} = v_{dc,k}^{ref}$. Then solving to i_{dg}^{ss}, i_{qg}^{ss} of (6.64) and (6.66), new variables $\hat{x}_{g,k} = [\hat{v}_{dc,k} \ \hat{i}_{dg,k} \ \hat{i}_{qg,k}]^T$,

$x_{g,k}^{ss} = [v_{dc,k}^{ss} \; i_{dg,k}^{ss} \; i_{qg,k}^{ss}]^T$ are defined, and the tracking error is written as

$$\varepsilon_{x_g,k} = \hat{x}_{g,k} - x_{g,k}^{ss}. \tag{6.67}$$

In order to simplify the controller synthesis, the DC Link identifier equations can be rewritten as

$$\hat{x}_{g,k+1} = \hat{f}_{x_g,k} + \hat{g}_{x_g,k} u_k, \tag{6.68}$$

where

$$\hat{f}_{x_g,k} = \begin{bmatrix} w_{11}S(v_{dc}) + w_{12}S(v_{dc})S(i_{qg}) + w_{13}i_{dg} \\ w_{21}S(i_{dg}) + w_{22}S(i_{qg}) + w_{23}S(v_{dc}) \\ w_{31}S(i_{qg}) + w_{32}S(i_{dg}) \end{bmatrix},$$

$$\hat{g}_{x_g,k} = \begin{bmatrix} 0 & 0 \\ w_{24,k} & 0 \\ 0 & w_{33,k} \end{bmatrix}.$$

Evaluating (6.67) at time $k + 1$, the tracking error dynamics are obtained as

$$\varepsilon_{x_g,k+1} = \hat{f}_{x_g,k} + \hat{g}_{x_g,k} u_k - x_{g,k+1}^{ss}. \tag{6.69}$$

Then the control signal $u_{g,k}$ is decomposed into two components as

$$u_{g,k} = u_{2,k} + u_{g,k}^*, \tag{6.70}$$

where

$$u_{2,k} = \hat{g}_{x_g,k}^{*-1}(x_{g,k+1}^{ss}), \tag{6.71}$$

with $\hat{g}_{x_g,k}^{*-1} = (\hat{g}_{x_g,k}^T \hat{g}_{x_g,k})^{-1} \hat{g}_{x_g,k}^T$. Then system (6.69), with (6.70) as input, results in

$$\varepsilon_{x_g,k+1} = \hat{f}_{x_g,k} + \hat{g}_{x_g,k} u_{g,k}^*. \tag{6.72}$$

System (6.72) is of the form (2.1), hence the proposed inverse optimal control law

$u^*_{g,k}$ using Theorem 4.1 becomes

$$u^*_{g,k} = -\frac{1}{2}(R_2(x_k) + \frac{1}{2}\hat{g}^T_{x_g,k}P_2\hat{g}_{x_g,k})^{-1}$$
$$\times \hat{g}^T_{x_g,k}P_2\hat{f}_{x_g,k}, \tag{6.73}$$

where for the grid side converter (GSC) controller, $R_2(x_k)$ and P_2 are the $R(x_k)$ and P matrices in (2.8), respectively.

6.3.2.1 Simulation Results

To evaluate the performance of the proposed controller, a simulation for a DC Link is developed. The DC Link parameters appear in Table 6.1. The simulation is performed in MATLAB/Simulink, with conditions:

- Simulation time: 7 seconds.
- Sampling time: $t_s = 0.5\ ms$.
- DC Link initial conditions: $v_{dc} = 0.01\ pu$, $i_{dg} = 0\ pu$, $i_{qg} = 0\ pu$.
- Identification input is a chirp signal, frequency range 0–60 Hz and amplitude 0.01 pu.
- The first 1 second is the identification; after that, the control signal is incepted at 1 second.
- The DC voltage reference is a constant signal at 0.5567 pu.
- Power factor reference is constant at 0.9.
- Load resistance R_L of 383.0579 pu connected to the capacitor in parallel scheme in order to simulate an unknown perturbation.

The behavior of the neural identifier is not presented in this section; it is shown in Figure 6.12 to Figure 6.14 in Section 6.2.2. At 1 second, the control signal is incepted. The performance of the DC Link controller is shown in Figure 6.22 to Figure 6.24. The DC voltage (Figure 6.22(a)) and power factor tracking (Figure 6.22(b)) are controlled to the reference. In this figure, we can see that the DC voltage reaches the reference quickly, and the electric power factor is kept constant at 0.9, with some numerical

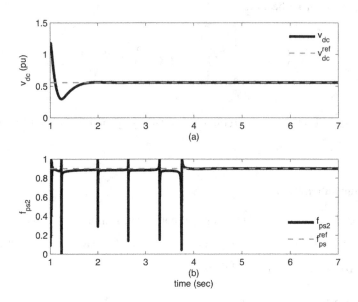

FIGURE 6.22   System outputs: (a) DC voltage (v_{dc}) and (b) step-up transformer power factor (f_{ps2}).

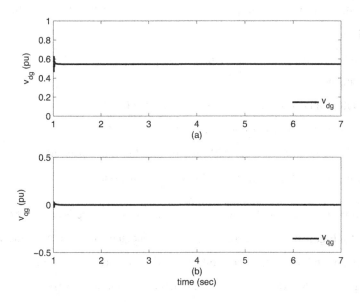

FIGURE 6.23   Control signals: (a) v_{dg} and (b) v_{qg}.

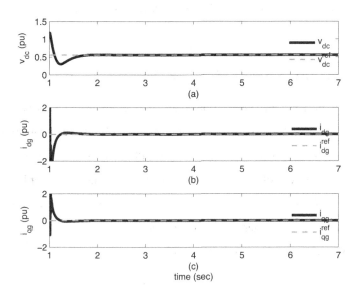

FIGURE 6.24   State variables: (a) DC voltage v_{dc}, (b) current i_{dg}, and (c) current i_{qg}.

discontinuities due to the electric power factor calculated as $f_{ps2} = \frac{P_g}{\sqrt{P_g^2 + Q_g^2}}$. Then
the discontinuities occur when P_g crosses zero and $Q_g \approx 0$. The control signals v_{dg}
and v_{qg} are bounded as shown in Figure 6.23. The state variables of the DC Link
are shown in Figure 6.24, where we can see that the state variables are stable and
bounded. The transient time is short and the resistance R_L is connected in a parallel
scheme to simulate the capacitor discharge by a resistive load.

6.4   IMPLEMENTATION ON A WIND ENERGY TESTBED

So far, the algorithms developed throughout this book have proven their performance
by means of simulations. The real-time implementation of control algorithms represents a great challenge because during the controller design, hypotheses are considered to facilitate the development and often in an implementation. Frequently in
real-time implementations, not all hypotheses can be fulfilled. There are also physical
limitations on the prototype that affects the implementation.

In this section, the real-time implementation of the designed algorithms is presented. Additionally, a comparison of the statistical information of the results is included.

6.4.1   REAL-TIME CONTROLLER PROGRAMING

The dSPACE[1] DS1104 signal acquisition board provides libraries that are compatible with the software MATLAB/Simulink. Additionally, the dSPACE company provides a monitoring software named *ControlDesk*, which allows one to monitor and interact in real time with the control algorithm loaded in the data acquisition board.

If the *ControlDesk* software is installed correctly, the DS1104 card libraries are also automatically loaded when MATLAB is loaded. It is verified in the command window of MATLAB, as shown in Figure 6.25.

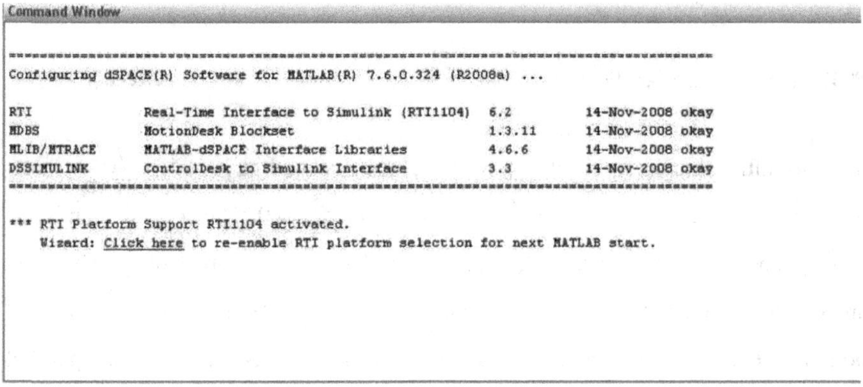

```
Command Window

================================================================================
Configuring dSPACE(R) Software for MATLAB(R) 7.6.0.324 (R2008a) ...

RTI              Real-Time Interface to Simulink (RTI1104)   6.2        14-Nov-2008 okay
MDBS             MotionDesk Blockset                         1.3.11     14-Nov-2008 okay
MLIB/MTRACE      MATLAB-dSPACE Interface Libraries           4.6.6      14-Nov-2008 okay
DSSIMULINK       ControlDesk to Simulink Interface           3.3        14-Nov-2008 okay
================================================================================

*** RTI Platform Support RTI1104 activated.
    Wizard: Click here to re-enable RTI platform selection for next MATLAB start.
```

FIGURE 6.25   Command window of MATLAB.

The *ControlDesk* software adds new blocks to Simulink, which are useful for the design of an algorithm using the data acquisition board hardware. The new blocks can be used as a standard block, which facilitates the incorporation of these libraries into a Simulink model, as shown in Figure 6.26. The dSPACE RTI1104 libraries can be used to access the analog-to-digital converters (ADC), as well as pulse width modulation (PWM) output ports.

---

[1]DS1104 R&D Controller Board of dSPACE GmbH.

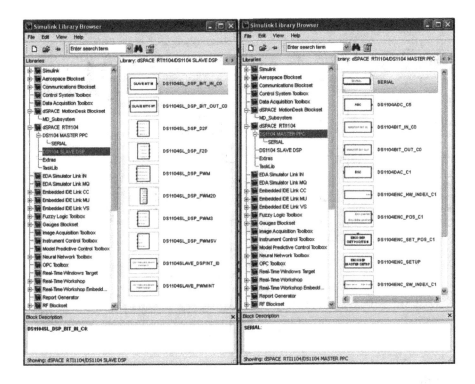

FIGURE 6.26 DS1104 Simulink libraries.

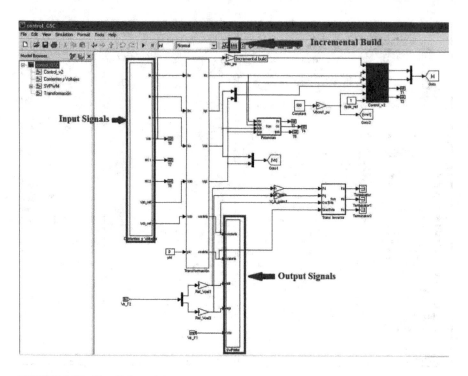

FIGURE 6.27   Simulink model.

The control algorithms previously designed in this book are implemented in a Simulink model, where the input signals are the ADC ports and the output signals are the PWM ports, as shown in Figure 6.27. Once the Simulink model is implemented and after quick configurations, the algorithm compilation is done automatically by pressing the button *Incremental Build* in the toolbar.

When the Simulink model is compiled, a code file with extension *.sdf* is generated. It is loaded directly to the DS1104 board, which executes the code in real time. The *ControlDesk* allows the monitoring and interaction of the variables within the algorithm in real time. The interaction of the user with the algorithm is performed through a fully customizable interface, which can be designed using the virtual instruments available in the *ControlDesk*. An interface to monitor the algorithm previously designed can be seen in Figure 6.28. The main advantage of using a dSPACE data acquisition board DS1104 is that it simplifies the real-time implementation of an algorithm programmed in Simulink.

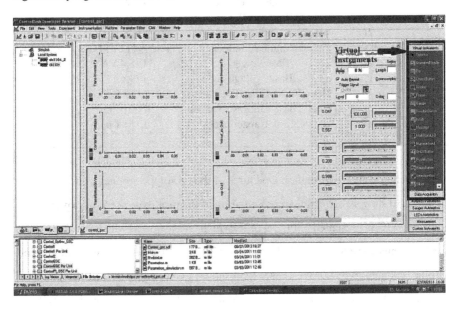

FIGURE 6.28   *ControlDesk* interface.

6.4.2   DOUBLY FED INDUCTION GENERATOR PROTOTYPE

In order to evaluate the performance of the proposed controller schemes, a low-power (1/4 HP) prototype is integrated. The complete doubly fed induction generator (DFIG) prototype is displayed in Figure 6.29, for which the nominal parameters appear in Table 6.1. This prototype includes four major parts: a 1/4 HP three-phase DFIG, a DC motor, two PWM units for the power stage, and a personal computer (PC) for supervising, which has the data acquisition board installed.

TABLE 6.1

Parameters of Doubly Fed Induction Generator Prototype

Symbol	Parameter	Value
X_m	Magnetizing Reactance	2.3175 pu
X_s	Stator Reactance	2.4308 pu
X_r	Rotor Reactance	2.4308 pu
r_s	Stator Windings Resistance	0.1609 pu
r_r	Rotor Windings Resistance	0.0502 pu
H	Angular Moment of Inertia	0.23 sec
ω_b	Base Angular Frequency	376.99112 rad/sec
P_b	Base Power	185.4 VA
V_b	Base Voltage	179.63 V
X_l	Three Phase Lines Reactance	0.0045 pu
r_g	Three Phase Lines Resistance	0.0014 pu
C	DC Link Capacitance	0.1854 pu

The scheme and corresponding pictures of the prototype are included as follows. Figure 6.30 shows a schematic representation of the prototype used for experiments. Just the DFIG is presented in Figure 6.31, which was acquired with Labvolt.[2] Figure 6.32 presents the DC motor (Baldor[3] 3/4 HP), which is used to emulate a wind turbine, coupled to the DFIG. Figure 6.33 shows a view of the PC and the DS1104[4] data acquisition board, which allows one to download applications directly from

---

[2] www.labvolt.com.

[3] www.baldor.com.

[4] DS1104 R&D Controller Board of dSPACE GmbH

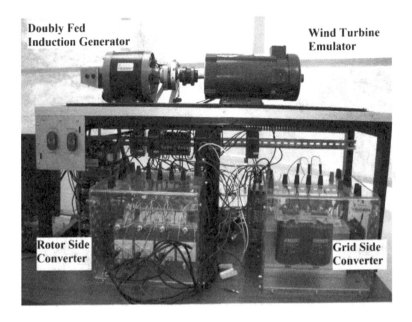

FIGURE 6.29   DFIG prototype description.

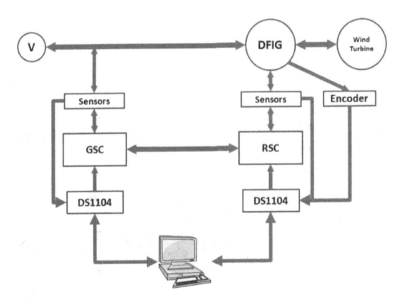

FIGURE 6.30   Prototype scheme.

FIGURE 6.31   1/4 HP DFIG.

FIGURE 6.32   Wind turbine emulated by a DC motor.

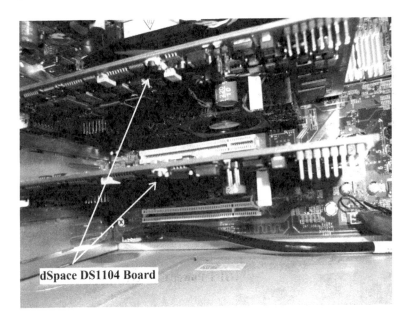

FIGURE 6.33   DS1104 data acquisition board.

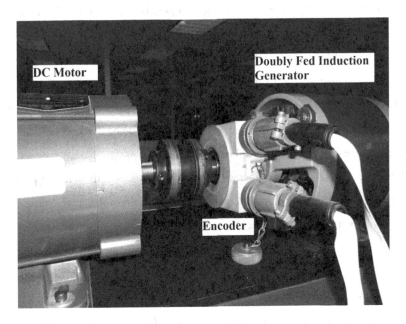

FIGURE 6.34   Encoder coupled between the DFIG and the DC motor.

FIGURE 6.35   PWM driver.

Simulink. Figure 6.34 portrays the encoder coupled between the DFIG and the DC motor to sense the mechanical rotor speed, and Figure 6.35 shows one of the PWM drivers. The connection of the DFIG prototype to the power system is done through a three-phase auto-transformer, which is displayed in Figure 6.36.

The control scheme implementation is performed using MATLAB/Simulink, with a DS1104 data acquisition board. The presence of unmodeled dynamics is one of the main challenges for a real-time implementation.

Table 6.2 shows the control schemes implemented successfully in the DFIG proto-type. In Sections 6.4.3, 6.4.4, and 6.4.5, the sliding mode control, neural sliding modes control, and neural inverse optimal control performance are presented, respectively. Each section presents the DFIG outputs and the DC Link outputs that are controlled for the proposed control scheme and their respective control signals as well. It is important that all controllers are tested under the same real-time conditions for the same output objectives. For all the implementations, the DC motor imposes a constant rotor speed of 0.97 *pu* as we can see in Figure 6.37.

FIGURE 6.36   Three-phase auto-transformer.

TABLE 6.2

Control Schemes Implemented in Real-Time Successfully

System	Sliding Modes	Inverse Optimal Control	Neural Sliding Modes	Neural Inverse Optimal Control
DFIG	OK	Fail	OK	OK
DC Link	OK	Fail	OK	OK

The real-time implementation of the inverse optimal control algorithm was not possible due to us knowing the nominal parameters of the prototype, which naturally vary during operation of the generator, and the controller depends strongly on knowledge of these parameters, which cannot be measured in real time. This situation was remedied by reformulating the inverse optimal control based on a neural model, as can be seen in Section 6.3.

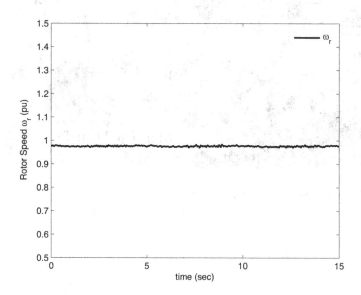

FIGURE 6.37   Rotor speed (ω_r) imposed by the DC motor.

6.4.3 SLIDING MODE REAL-TIME RESULTS

In this section, the real-time results of the sliding mode controllers are presented. The real-time implementation conditions are:

- Capture time: 15 seconds.
- Sampling time: $t_s = 0.5\ ms$.
- The electric torque reference τ_e^{ref} is a senoidal signal centered at 0.5 pu with amplitude of 0.2 pu and 0.2 Hz.
- Power factor reference f_{ps1}^{ref} is constant at 1.0.
- The DC voltage reference v_{dc}^{ref} is a constant signal at 0.5567 pu.
- Power factor reference f_{ps2}^{ref} is constant at 1.0.

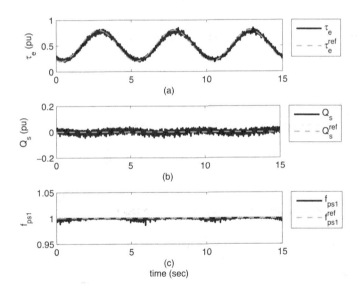

FIGURE 6.38 DFIG outputs with sliding modes: (a) electric torque (τ_e) tracking, (b) reactive power (Q_s) tracking, and (c) electric power factor (f_{ps1}).

Figure 6.38 presents (a) the electric torque (τ_e) tracking, (b) the reactive power (Q_s) tracking and (c) the electric power factor. In this figure, it can be seen that the tracking for electric torque is achieved despite the reference signal being time variant. In Figure 6.38(b), it can be seen that the reactive power kept almost constant at 0 pu is

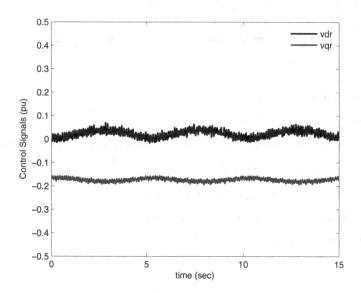

FIGURE 6.39   DFIG control signals v_{dr} and v_{qr} for sliding modes.

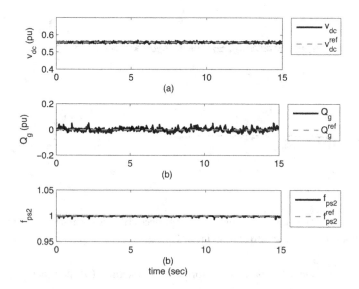

FIGURE 6.40   DC Link outputs with sliding modes: (a) DC voltage (v_{dc}) tracking, (b) reactive power (Q_g) tracking, and (c) electric power factor (f_{ps2}) in the step-up transformer.

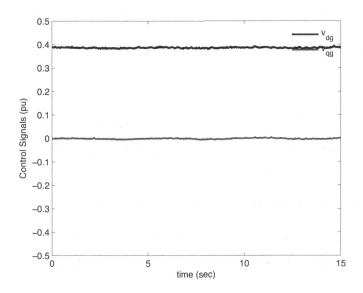

FIGURE 6.41 DC Link control signals v_{dg} and v_{qg} for sliding modes.

slightly affected by the dynamics of the electric torque reference; even so, the power factor (f_{ps1}) is not affected significantly as we can see in Figure 6.38(c). The control signals are bounded and these are shown in Figure 6.39.

The performance of the DC Link controller is shown in Figure 6.40 and Figure 6.41. The DC voltage (v_{dc}) (Figure 6.40(a)) and reactive power (Q_g) (Figure 6.40(b)) are controlled to the reference. In this figure, we can see that the DC voltage remained in the reference, and the reactive power (Q_g) kept the average value in the reference; it presents small variations due to noise in the measurement of currents and voltages, including the effect of the switching of the insulated-gate bipolar transistor (IGBT), however, the control objective is achieved. The electric power factor (f_{ps2}) is kept very close to 1. The control signals v_{dg} and v_{qg} are bounded, as shown in Figure 6.41.

Quantitative measures of the performance of this real-time implementation are shown in Table 6.3, where $\varepsilon_{\tau_e} = \tau_e - \tau_e^{ref}$, $\varepsilon_{Q_s} = Q_s - Q_s^{ref}$, $\varepsilon_{f_{ps1}} = f_{ps1} - f_{ps1}^{ref}$, $\varepsilon_{v_{dc}} = v_{dc} - v_{dc}^{ref}$, $\varepsilon_{Q_g} = Q_g - Q_g^{ref}$, and $\varepsilon_{f_{ps2}} = f_{ps2} - f_{ps2}^{ref}$.

TABLE 6.3

Statistical Measures of Real-Time Implementation Results of the Sliding Modes Controller

Measure	ε_{τ_e}	ε_{Q_s}	$\varepsilon_{f_{ps1}}$	$\varepsilon_{v_{dc}}$	ε_{Q_g}	$\varepsilon_{f_{ps2}}$
MEAN	1.0792e-005	3.3333e-005	-6.8908e-004	6.8583e-006	3.7931e-004	-3.9046e-004
STD	0.0424	0.0146	0.0012	0.0024	0.0128	6.1217e-004
MSE	0.0018	2.1318e-004	1.8404e-006	5.7453e-006	1.6297e-004	5.2719e-007

6.4.4 NEURAL SLIDING MODE REAL-TIME RESULTS

In this section, the real-time results of the neural sliding mode controllers designed in Section 6.2 are presented. The real-time implementation conditions are:

- Capture time: 15 seconds.
- Sampling time: $t_s = 0.5$ *ms*.
- The electric torque reference τ_e^{ref} is a senoidal signal centered at 0.5 *pu* with amplitude of 0.2 *pu* and 0.2 *Hz*.
- Power factor reference f_{ps1}^{ref} is constant at 1.0.
- The DC voltage reference v_{dc}^{ref} is a constant signal at 0.5567 *pu*.
- Power factor reference f_{ps2}^{ref} is constant at 1.0.

Figure 6.42 presents (a) the electric torque (τ_e) tracking, (b) the reactive power (Q_s) tracking, and (c) the electric power factor. In this figure, it can be seen that the tracking for electric torque is achieved. In Figure 6.42(b), it can be seen that the reactive power kept almost constant at 0 *pu* was slightly affected by the dynamics of the electric torque reference, similar to the sliding mode controller; and the power factor (f_{ps1}) variations are more evident compared with the ones of the previous controller, as can be see in Figure 6.42(c). The control signals are bounded and these are shown in Figure 6.43.

The performance of the DC Link neural controller is shown in Figure 6.44. The DC voltage (v_{dc}) is shown in Figure 6.44(a), where we can see that the DC voltage

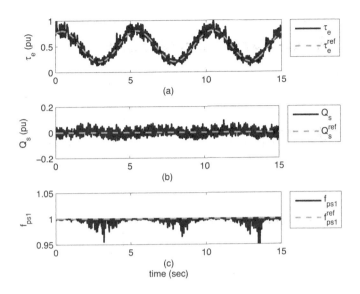

FIGURE 6.42 DFIG outputs with neural sliding modes: (a) electric torque (τ_e) tracking, (b) reactive power (Q_s) tracking, and (c) electric power factor (f_{ps1}).

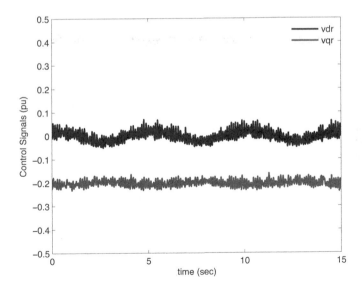

FIGURE 6.43 DFIG control signals v_{dr} and v_{qr} for neural sliding modes.

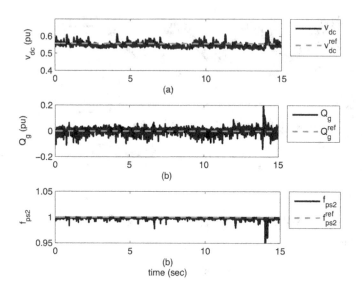

FIGURE 6.44   DC Link outputs with neural sliding modes: (a) DC voltage (v_{dc}) tracking, (b) reactive power (Q_g) tracking, and (c) electric power factor (f_{ps2}) in the step-up transformer.

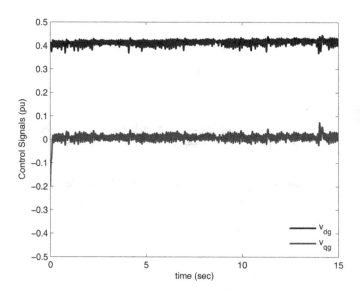

FIGURE 6.45   DC Link control signals v_{dg} and v_{qg} for neural sliding modes.

TABLE 6.4

Statistical Measures of Real-Time Implementation Results of the Neural Sliding Modes Controller

Measure	ε_{τ_e}	ε_{Q_s}	$\varepsilon_{f_{ps1}}$	$\varepsilon_{v_{dc}}$	ε_{Q_g}	$\varepsilon_{f_{ps2}}$
MEAN	0.0060	-8.5726e-005	-0.0025	-0.0049	0.0063	-0.0024
STD	0.0684	0.0260	0.0045	0.0175	0.0296	0.0118
MSE	0.0047	6.7426e-004	2.6214e-005	3.3153e-004	9.1429e-004	1.4373e-004

remained close to the reference, but it makes variations more evident. The mean value of the reactive power (Q_g) is in the reference, but it presents variations due to noise in the measurement of currents and voltages, including the effect of the switching of the IGBT, similar to the previous controller, however, the control objective is achieved. It is evident that the electric power factor (f_{ps2}) is kept close to 1 even though there are slight variations in the reactive power. The control signals v_{dg} and v_{qg} are bounded, as shown in Figure 6.45.

Quantitative measures of the performance of this real-time implementation are shown in Table 6.4.

6.4.5 NEURAL INVERSE OPTIMAL CONTROL REAL-TIME RESULTS

In this section, the real-time results of the neural inverse optimal controllers designed in Section 6.3 are presented. The real-time implementation conditions are:

- Capture time: 15 seconds.
- Sampling time: $t_s = 0.5$ ms.
- The electric torque reference τ_e^{ref} is a senoidal signal centered at 0.5 pu with amplitude of 0.2 pu and 0.2 Hz.
- Power factor reference f_{ps1}^{ref} is constant at 1.0.
- The DC voltage reference v_{dc}^{ref} is a constant signal at 0.5567 pu.
- Power factor reference f_{ps2}^{ref} is constant at 1.0.

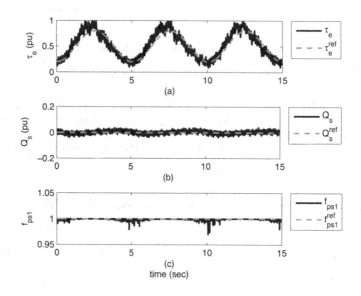

FIGURE 6.46   DFIG outputs with neural inverse optimal: (a) electric torque (τ_e) tracking, (b) reactive power (Q_s) tracking, and (c) electric power factor (f_{ps1}).

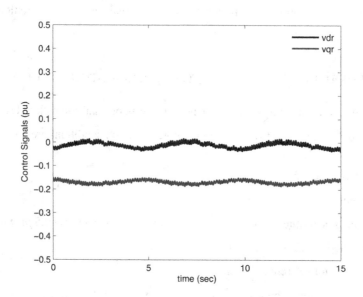

FIGURE 6.47   DFIG control signals v_{dr} and v_{qr} for neural inverse optimal.

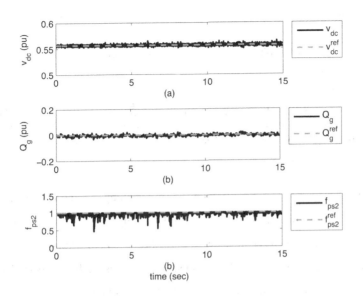

FIGURE 6.48   DC Link outputs with neural inverse optimal: (a) DC voltage (v_{dc}) tracking, (b) reactive power (Q_g) tracking, and (c) electric power factor (f_{ps2}) in the step-up transformer.

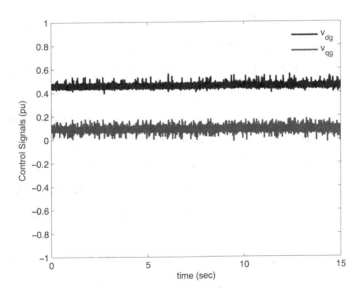

FIGURE 6.49   DC Link control signals v_{dg} and v_{qg} for neural inverse optimal.

TABLE 6.5

Statistical Measures of Real-Time Implementation Results of the Neural Inverse Optimal Controller

Measure	ε_{τ_e}	ε_{Q_s}	$\varepsilon_{f_{ps1}}$	$\varepsilon_{v_{dc}}$	ε_{Q_g}	$\varepsilon_{f_{ps2}}$
MEAN	0.0160	5.5803e-005	-8.0718e-004	-6.6140e-006	-0.0038	-0.0405
STD	0.0962	0.0139	0.0017	0.0010	0.0070	0.0551
MSE	0.0095	1.9200e-004	3.4869e-006	1.0475e-006	6.2826e-005	0.0047

Figure 6.46 presents (a) the electric torque (τ_e) tracking, (b) the reactive power (Q_s) tracking, and (c) the electric power factor. In this figure, it can be seen that the tracking for electric torque is achieved with a small error. In Figure 6.46(b), it can be seen that the reactive power kept almost constant at 0 pu is slightly affected by the dynamics of the electric torque reference similar to the sliding mode controller; and the power factor (f_{ps1}) variations are small like the sliding modes controller, as can be seen in Figure 6.46(c). The control signals are bounded and these are shown in Figure 6.47.

The performances of the DC Link neural controller and the respective control signals are shown in Figure 6.48 and Figure 6.49, respectively. The DC voltage (v_{dc}) is shown in Figure 6.48(a), where we can see that the DC voltage tracking is very good. The reactive power (Q_g) tracking is achieved successfully, but it presents variations due to noise in the measurement of currents and voltages, including the effect of the switching of the IGBT, similar to previous controllers; however, the control objective is achieved. We can see that the electric power factor (f_{ps2}) is kept close to the reference at 1.0. The control signals v_{dg} and v_{qg} are bounded, as shown in Figure 6.49.

Quantitative measures of the performance of this real-time implementation are shown in Table 6.5.

Based on the above real-time implementation results, Table 6.6 presents the statistical measures to determine the best controller implemented.

As we can see from Table 6.6, the controllers with less tracking error are developed using discrete time sliding mode and inverse optimal control with neural networks.

TABLE 6.6
Statistical Measures

Control Scheme	Measure	ε_{τ_e}	ε_{Q_s}	$\varepsilon_{f_{ps1}}$	$\varepsilon_{v_{dc}}$	ε_{Q_g}	$\varepsilon_{f_{ps2}}$
Sliding Modes	MEAN	**1.08e-5**	3.33e-5	**-6.89e-4**	6.86e-6	**3.79e-4**	-3.90e-4
	STD	**0.0424**	0.0146	**0.0012**	0.0024	**0.0128**	6.12e-4
	MSE	**0.0018**	2.13e-4	**1.84e-6**	5.74e-6	**1.63e-4**	5.27e-7
Neural Sliding Modes	MEAN	0.0060	-8.57e-5	-0.0025	-0.0049	0.0063	-0.0024
	STD	0.0684	0.0260	0.0045	0.0175	0.0296	0.0118
	MSE	0.0047	6.74e-4	2.62e-5	3.31e-4	9.14e-4	1.44e-4
Neural Inverse Optimal	MEAN	0.0160	**5.58e-5**	-8.07e-4	**-6.61e-6**	-0.0038	-0.0405
	STD	0.0962	**0.0139**	0.0017	**0.0010**	0.0070	0.0551
	MSE	0.0095	**1.92e-4**	3.49e-6	**1.05e-6**	6.28e-5	0.0047

Although the neural sliding mode controller does not have the lowest tracking error, the convergence time to the reference is smaller than the one of the other controllers. The real-time results that make this fact obvious are not presented in this book due to space restrictions.

The main advantage of neural inverse optimal control is that this algorithm has smoother control signals, which can be seen by comparing the control signals in Figure 6.39, Figure 6.43, and Figure 6.47 for the DFIG, and the respective Figure 6.41, Figure 6.45 and Figure 6.49 for the DC Link.

All results presented in this chapter validate the effectiveness of the algorithms developed in this book.

6.5 CONCLUSIONS

In this chapter, the developed controllers are applied, based on the respective neural model for a DFIG. The control schemes use dSPACE DS1104 Controller Board. Simulation and real-time implementation of the schemes proposed are presented, validating that the theoretical results are achieved for a DFIG.

REFERENCES

1. P. W. Carlin, A. S. Laxson, and E. B. Muljadi. The history and state of the art of variable-speed wind turbine technology. *Wind Energy*, 6(2):129-159, 2003.

2. J. Hu, H. Nian, B. Hu, Y. He, and Z.Q. Zhu. Direct active and reactive power regulation of DFIG using sliding-mode control approach. *IEEE Transactions on Energy Conversion*, 25(4):1028-1039, 2010.

3. J. Lopez, P. Sanchis, X. Roboam, and L. Marroyo. Dynamic behavior of the doubly fed induction generator during three-phase voltage dips. *IEEE Transactions on Energy Conversion*, 22(3):709-717, 2007.

4. A. G. Loukianov. Nonlinear block control with sliding modes. *Automation and Remote Control*, 57(7):916–933, 1998.

5. A. Monroy, L. Alvarez-Icaza, and G. Espinosa-Pérez. Passivity-based control for variable speed constant frequency operation of a DFIG wind turbine. *International Journal of Control*, 81(9):1399-1407, 2008.

6. S. Muller, M. Deicke, and R. W. De Doncker. Doubly fed induction generator systems for wind turbines. *IEEE Industry Applications Magazine*, 8(3), 26-33, 2002.

7. R. Pena, J. C. Clare, and G. M. Asher. Doubly fed induction generator using back-to-back PWM converters and its application to variable-speed wind-energy generation. *IEE Proceedings-Electric Power Applications*, 143(3):231-241, 1996.

8. R. Ruiz-Cruz, E. N. Sanchez, F. Ornelas-Tellez, A. G. Loukianov, and R. G. Harley. Particle swarm optimization for discrete-time inverse optimal control of a doubly fed induction generator. *IEEE Transactions on Cybernetics*, 43(6):168-1709, 2013.

9. E. N. Sanchez, and R. Ruiz-Cruz. *Doubly Fed Induction Generators: Control for Wind Energy*. CRC Press, 2016.

10. V. Utkin, J. Guldner, and M. Shijun. *Sliding Mode Control in Electro-mechanical Systems*. Automation and Control Engineering. Taylor & Francis, 1999.

11. F. Wu, and X.P. Zhang, and P. Ju, and M. Sterling. Decentralized nonlinear

control of wind turbine with doubly fed induction generator. *IEEE Transactions on Power Systems*, 23(2):613-621, 2008.

7 Conclusions

In this book, based on the neural networks, sliding mode, and inverse optimal control techniques, two novel methodologies to synthesize robust controllers for a class of MIMO discrete-time nonlinear uncertain systems are proposed, as follows: The first control scheme is developed using a recurrent high order neural network which enables identification of the plant model. A strategy to avoid specific adaptive weights zero-crossing and conserve the identifier controllability property is proposed. Based on this neural identifier and applying the discrete-time block control approach, a nonlinear sliding manifold with a desired asymptotically stable motion is formulated. Using a Lyapunov functions approach, a discrete-time sliding mode control which ensures that the sliding manifolds are attractive, is introduced.

Then a discrete-time inverse optimal control scheme is developed, which achieves stabilization and trajectory tracking for nonlinear systems and is inverse optimal in the sense that, a posteriori, it minimizes a cost functional. To avoid the Hamilton–Jacobi–Bellman equation solution, we propose a discrete-time quadratic control Lyapunov function (CLF). Furthermore, a robust inverse optimal control is established in order to guarantee stability for nonlinear systems, which are affected by internal and/or external disturbances. We use discrete-time recurrent neural networks to model uncertain nonlinear systems; thus, an explicit knowledge of the plant is not necessary. The proposed approach is successfully applied to implement a robust controller based on a recurrent high order neural network identifier and inverse optimality. By means of simulations, it can be seen that the required goal is achieved, i.e., the proposed controller maintains stability of the plant with unknown parameters. For neural network training, an on-line extended Kalman filter is performed.

Both the first and second control schemes require only plant model structure knowledge, but the plant state vector must be available for measurement. In the case when only the plant output is measured, an observer needs to be designed. Simulation and real-time implementation of the schemes proposed in this book are presented, validating the theoretical results, using two benchmarks, the first for a three-phase induction motor and the second one for a DFIG.

The two developed neural control schemes are implemented in real time for two kinds of very useful electric machines: induction motors and double fed induction generators. The experimental results illustrate the robustness of the designed controllers with respect to plant parameters variations and external disturbances.

A. DFIG and DC Link

Mathematical Model

A.1 DFIG MATHEMATICAL MODEL

The induction machine is used in a wide variety of applications as a means of converting electric power to mechanical work or the other way around. The voltage

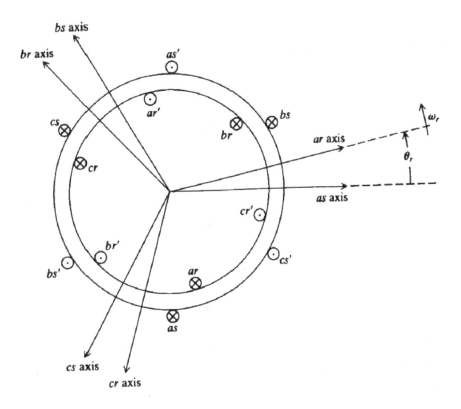

FIGURE A.1   Two-pole, 3-phase, elementary induction machine.

equations for the elementary induction machine shown in Figure A.1 and Figure A.2

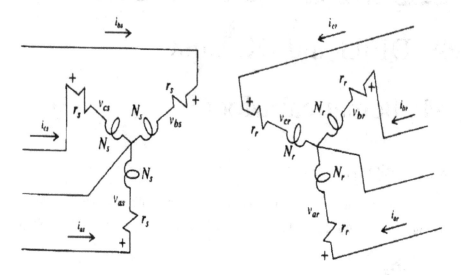

FIGURE A.2   Direction currents induction motor convention.

are:

$$v_{as} = r_s i_{as} + \frac{d\lambda_{as}}{dt},$$

$$v_{bs} = r_s i_{bs} + \frac{d\lambda_{bs}}{dt},$$

$$v_{cs} = r_s i_{cs} + \frac{d\lambda_{cs}}{dt}, \qquad \text{(A.1)}$$

$$v_{ar} = r_r i_{ar} + \frac{d\lambda_{ar}}{dt},$$

$$v_{br} = r_r i_{br} + \frac{d\lambda_{br}}{dt},$$

$$v_{cr} = r_r i_{cr} + \frac{d\lambda_{cr}}{dt},$$

where v_{as}, v_{bs}, v_{cs} are the stator voltages in the a, b, c axis, respectively; i_{as}, i_{bs}, i_{cs} are the stator currents; v_{ar}, v_{br}, v_{cr} are the rotor voltages; i_{ar}, i_{br}, i_{cr} are the rotor currents; r_s is the resistance of the stator winding; r_r is the resistance of the rotor winding. The flux linkages are expressed as

$$\lambda_{as} = L_{asas}i_{as} + L_{asbs}i_{bs} + L_{ascs}i_{cs} + L_{asar}i_{ar} + L_{asbr}i_{br} + L_{ascr}i_{cr},$$

$$\lambda_{bs} = L_{bsas}i_{as} + L_{bsbs}i_{bs} + L_{bscs}i_{cs} + L_{bsar}i_{ar} + L_{bsbr}i_{br} + L_{bscr}i_{cr},$$

$$\lambda_{cs} = L_{csas}i_{as} + L_{csbs}i_{bs} + L_{cscs}i_{cs} + L_{csar}i_{ar} + L_{csbr}i_{br} + L_{cscr}i_{cr}, \qquad (A.2)$$

$$\lambda_{ar} = L_{aras}i_{as} + L_{arbs}i_{bs} + L_{arcs}i_{cs} + L_{arar}i_{ar} + L_{arbr}i_{br} + L_{arcr}i_{cr},$$

$$\lambda_{br} = L_{bras}i_{as} + L_{brbs}i_{bs} + L_{brcs}i_{cs} + L_{brar}i_{ar} + L_{brbr}i_{br} + L_{brcr}i_{cr},$$

$$\lambda_{cr} = L_{cras}i_{as} + L_{crbs}i_{bs} + L_{crcs}i_{cs} + L_{crar}i_{ar} + L_{crbr}i_{br} + L_{crcr}i_{cr},$$

The winding inductances of the induction machine may be expressed from the inductance relationships given for the salient-pole synchronous machine. In the case of the induction machine, the air gap is uniform. All stator self-inductances are equal; that is, $L_{asas} = L_{bsbs} = L_{cscs}$ with

$$L_{asas} = L_{ls} + L_{ms}, \qquad (A.3)$$

where L_{ms} is the stator magnetizing inductance. Likewise all stator-to-stator mutual inductances are the same.

$$L_{asbs} = L_{bscs} = L_{csas} = -\frac{L_{ms}}{2}. \qquad (A.4)$$

In Figure A.1, the ar axis is displaced with as axes at an angle θ_r. Similarly, the axes br, cr are displaced with the axes bs, cs, respectively, at the same angle θ_r. So the stator-to-rotor mutual inductances are defined as

$$L_{asar} = L_{bsbr} = L_{cscr} = \frac{N_r}{N_s} L_{ms} cos(\theta_r). \qquad (A.5)$$

The angle between the axis as and br is $\theta_r + \frac{2\pi}{3}$; then

$$L_{asbr} = L_{bscr} = L_{csar} = \frac{N_r}{N_s} L_{ms} cos(\theta_r + \frac{2\pi}{3}). \qquad (A.6)$$

The as axis is displaced with cr at an angle $\theta_r - \frac{2\pi}{3}$; then

$$L_{ascr} = L_{bsar} = L_{csbr} = \frac{N_r}{N_s} L_{ms} \cos\left(\theta_r - \frac{2\pi}{3}\right). \tag{A.7}$$

All rotor-to-rotor mutual inductances are the same and are defined as

$$L_{arar} = L_{brbr} = L_{crbr} = L_{lr} + \left(\frac{N_r}{N_s}\right)^2 L_{ms}, \tag{A.8}$$

where L_{lr} is the rotor leakage inductance. Finally, the mutual inductances between ar and br, br and cr, and cr and ar are defined in terms of the stator mutual inductances as

$$L_{arbr} = L_{brcr} = L_{crar} = -\left(\frac{N_r}{N_s}\right)^2 \frac{L_{ms}}{2}. \tag{A.9}$$

In order to simplify the handling of Equations (A.1) and (A.2), it can be rewritten in a matrix form as

$$
\begin{aligned}
v_{abcs} &= R_s i_{abcs} + \frac{d\lambda_{abcs}}{dt}, \\
v_{abcr} &= R_r i_{abcr} + \frac{d\lambda_{abcr}}{dt}, \\
\lambda_{abcs} &= L_{ss} i_{abcs} + L_{sr} i_{abcr}, \\
\lambda_{abcr} &= L_{sr}^T i_{abcs} + L_{rr} i_{abcr},
\end{aligned}
\tag{A.10}
$$

where

$$f_{abcx} = \begin{bmatrix} f_{ax} \\ f_{bx} \\ f_{cx} \end{bmatrix}. \tag{A.11}$$

The symbol f is used to represent the voltages, currents, and coupling fluxes; the x subscript is used to refer to the stator or rotor. In addition

$$R_s = \begin{bmatrix} r_s & 0 & 0 \\ 0 & r_s & 0 \\ 0 & 0 & r_s \end{bmatrix}, R_r = \begin{bmatrix} r_r & 0 & 0 \\ 0 & r_r & 0 \\ 0 & 0 & r_r \end{bmatrix},$$

$$L_{ss} = \begin{bmatrix} L_{ls} + L_{ms} & -\frac{1}{2}L_{ms} & -\frac{1}{2}L_{ms} \\ -\frac{1}{2}L_{ms} & L_{ls} + L_{ms} & -\frac{1}{2}L_{ms} \\ -\frac{1}{2}L_{ms} & -\frac{1}{2}L_{ms} & L_{ls} + L_{ms} \end{bmatrix},$$

$$L_{rr} = \begin{bmatrix} L_{lr} + L_{mr} & -\frac{1}{2}L_{mr} & -\frac{1}{2}L_{mr} \\ -\frac{1}{2}L_{mr} & L_{lr} + L_{mr} & -\frac{1}{2}L_{mr} \\ -\frac{1}{2}L_{mr} & -\frac{1}{2}L_{mr} & L_{lr} + L_{mr} \end{bmatrix},$$

$$L_{sr} = \begin{bmatrix} L_{sr}\cos(\theta_r) & L_{sr}\cos(\theta_r + \frac{2\pi}{3}) & L_{sr}\cos(\theta_r - \frac{2\pi}{3}) \\ L_{sr}\cos(\theta_r - \frac{2\pi}{3}) & L_{sr}\cos(\theta_r) & L_{sr}\cos(\theta_r + \frac{2\pi}{3}) \\ L_{sr}\cos(\theta_r + \frac{2\pi}{3}) & L_{sr}\cos(\theta_r - \frac{2\pi}{3}) & L_{sr}\cos(\theta_r) \end{bmatrix},$$

where

$$L_{mr} = \left(\frac{N_r}{N_s}^2\right)L_{ms},$$

$$L_{sr} = \frac{N_r}{N_s}L_{ms}.$$

When the voltage equations are expressed as (A.10), it is convenient to refer all rotor variables to stator side using the tip ratio N_s/N_r; then

$$i'_{abcr} = \frac{N_r}{N_s}i_{abcr}, \tag{A.12}$$

$$\lambda'_{abcr} = \frac{N_s}{N_r}\lambda_{abcr}, \tag{A.13}$$

$$v'_{abcr} = \frac{N_s}{N_r}v_{abcr}, \tag{A.14}$$

$$r'_r = \left(\frac{N_s}{N_r}\right)^2 r_r, \tag{A.15}$$

$$L'_{lr} = \left(\frac{N_s}{N_r}\right)^2 L_{lr}. \tag{A.16}$$

Using the Equations (A.12) to (A.16), the equation system (A.10) can be rewritten as

$$
\begin{aligned}
v_{abcs} &= R_s i_{abcs} + \frac{d\lambda_{abcs}}{dt}, \\
v'_{abcr} &= R'_r i'_{abcr} + \frac{d\lambda'_{abcr}}{dt}, \\
\lambda_{abcs} &= L_{ss} i_{abcs} + L'_{sr} i'_{abcr}, \\
\lambda'_{abcr} &= L'_{sr}{}^T i_{abcs} + L'_{rr} i'_{abcr},
\end{aligned}
\tag{A.17}
$$

where

$$R'_r = \begin{bmatrix} r'_r & 0 & 0 \\ 0 & r'_r & 0 \\ 0 & 0 & r'_r \end{bmatrix},$$

$$L'_{rr} = \begin{bmatrix} L'_{lr} + L_{ms} & -\frac{1}{2}L_{ms} & -\frac{1}{2}L_{ms} \\ -\frac{1}{2}L_{ms} & L'_{lr} + L_{ms} & -\frac{1}{2}L_{ms} \\ -\frac{1}{2}L_{ms} & -\frac{1}{2}L_{ms} & L'_{lr} + L_{ms} \end{bmatrix},$$

$$L'_{sr} = \begin{bmatrix} L_{ms}\cos(\theta_r) & L_{ms}\cos(\theta_r + \frac{2\pi}{3}) & L_{ms}\cos(\theta_r - \frac{2\pi}{3}) \\ L_{ms}\cos(\theta_r - \frac{2\pi}{3}) & L_{ms}\cos(\theta_r) & L_{ms}\cos(\theta_r + \frac{2\pi}{3}) \\ L_{ms}\cos(\theta_r + \frac{2\pi}{3}) & L_{ms}\cos(\theta_r - \frac{2\pi}{3}) & L_{ms}\cos(\theta_r) \end{bmatrix},$$

The doubly fed induction generator (DFIG) mathematical model (A.17) considers a direction currents convention as the motor shown in Figure A.2. In this book, the direction currents convention selected is shown in Figure A.3. Then the DFIG mathematical model (A.17) is rewritten as

$$v_{abcs} = -R_s i_{abcs} + \frac{d\lambda_{abcs}}{dt},$$

$$v'_{abcr} = R'_r i'_{abcr} + \frac{d\lambda'_{abcr}}{dt}, \qquad \text{(A.18)}$$

$$\lambda_{abcs} = -L_{ss} i_{abcs} + L'_{sr} i'_{abcr},$$

$$\lambda'_{abcr} = -L'^T_{sr} i_{abcs} + L'_{rr} i'_{abcr}.$$

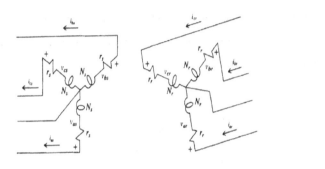

FIGURE A.3   Direction currents DFIG convention.

A.1.1   VARIABLES TRANSFORMATION REFERRED TO A REFERENCE FRAME FIXED IN THE ROTOR

The voltage equations that describe the performance of induction machines are functions of the rotor speed, whereupon the coefficients of the differential equations that describe the behavior of these machines are time-varying except when the rotor is stalled. A change of variables is often used to reduce the complexity of these differential equations. A general transformation refers machine variables to a frame of reference that rotates at an arbitrary angular velocity. All known real transformations are obtained from this transformation by simply assigning the speed of the rotation of the reference frame.

The time-varying inductances of a synchronous machine are eliminated only if the reference frame is fixed in the rotor, which is called *dq* transformation.

A change of variables that formulates a transformation of the three phase variables of the stator circuits to the arbitrary reference frame is expressed by

$$
K_s = \frac{2}{3}
\begin{bmatrix}
\cos\theta & \cos\left(\theta - \frac{2\pi}{3}\right) & \cos\left(\theta + \frac{2\pi}{3}\right) \\
-\sin\theta & -\sin\left(\theta - \frac{2\pi}{3}\right) & -\sin\left(\theta + \frac{2\pi}{3}\right) \\
\frac{1}{2} & \frac{1}{2} & \frac{1}{2}
\end{bmatrix},
\tag{A.19}
$$

where $\omega = \frac{d\theta}{dt}$ is the angular speed of the reference frame. However, in the analysis of induction machines it is also desirable to transform the variables associated with the symmetrical rotor windings to the arbitrary reference frame. A change of variables that formulates a transformation of the three phase variables of the rotor circuits to the arbitrary reference frame is:

$$
K_r = \frac{2}{3}
\begin{bmatrix}
\cos\beta & \cos\left(\beta - \frac{2\pi}{3}\right) & \cos\left(\beta + \frac{2\pi}{3}\right) \\
-\sin\beta & -\sin\left(\beta - \frac{2\pi}{3}\right) & \cos\left(\beta + \frac{2\pi}{3}\right) \\
\frac{1}{2} & \frac{1}{2} & \frac{1}{2}
\end{bmatrix},
\tag{A.20}
$$

with

$$
\beta = \theta - \theta_r.
$$

So, the stator and rotor variables transformed using transformations (A.19) and (A.20) are defined, respectively, as

$$
f_{dq0s} = K_s f_{abcs},
\tag{A.21}
$$

$$
f_{dq0r} = K_r f_{abcr},
\tag{A.22}
$$

where the symbol f is used to refer to each DFIG variable.

Applying transformations (A.19) and (A.20) to the DFIG Equations (A.18), the following equations are obtained:

$$K_s^{-1}v_{dq0s} = -R_s K_s^{-1} i_{dq0s} + \frac{d}{dt}\left[K_s^{-1}\lambda_{dq0s}\right],$$

$$K_r^{-1}v'_{dq0r} = R'_r K_r^{-1} i'_{dq0r} + \frac{d}{dt}\left[K_r^{-1}\lambda'_{dq0r}\right], \qquad \text{(A.23)}$$

$$K_s^{-1}\lambda_{dq0s} = -L_{ss} K_s^{-1} i_{dq0s} + L'_{sr} K_r^{-1} i'_{dq0r},$$

$$K_r^{-1}\lambda'_{dq0r} = -L'^{T}_{sr} K_s^{-1} i_{dq0s} + L'_{rr} K_r^{-1} i'_{dq0r},$$

where

$$K_s^{-1} = \begin{bmatrix} \cos\theta & -\sin\theta & 1 \\ \cos(\theta - \frac{2\pi}{3}) & -\sin(\theta - \frac{2\pi}{3}) & 1 \\ \cos(\theta + \frac{2\pi}{3}) & -\sin(\theta + \frac{2\pi}{3}) & 1 \end{bmatrix}, \qquad \text{(A.24)}$$

$$K_r^{-1} = \begin{bmatrix} \cos\beta & -\sin\beta & 1 \\ \cos(\beta - \frac{2\pi}{3}) & -\sin(\beta - \frac{2\pi}{3}) & 1 \\ \cos(\beta + \frac{2\pi}{3}) & -\sin(\beta + \frac{2\pi}{3}) & 1 \end{bmatrix}, \qquad \text{(A.25)}$$

with $\beta = \theta - \theta_r$. Developing (A.23) and reordering terms,

$$v_{dq0s} = -K_s R_s K_s^{-1} i_{dq0s} + K_s \frac{d}{dt}\left[K_s^{-1}\right]\lambda_{dq0s} + \frac{d}{dt}\left[\lambda_{dq0s}\right],$$

$$v'_{dq0r} = K_r R'_r K_r^{-1} i'_{dq0r} + K_r \frac{d}{dt}\left[K_r^{-1}\right]\lambda'_{dq0r} + \frac{d}{dt}\left[\lambda'_{dq0r}\right], \qquad \text{(A.26)}$$

$$\lambda_{dq0s} = -K_s L_{ss} K_s^{-1} i_{dq0s} + K_s L'_{sr} K_r^{-1} i'_{dq0r},$$

$$\lambda'_{dq0r} = -K_r L'^{T}_{sr} K_s^{-1} i_{dq0s} + K_r L'_{rr} K_r^{-1} i'_{dq0r},$$

where

$$K_s R_s K_s^{-1} = R_s,$$

$$K_r R'_r K_r^{-1} = R'_r,$$

$$K_s \frac{d}{dt}\left[K_s^{-1}\right] = \begin{bmatrix} 0 & -\omega_s & 0 \\ \omega_s & 0 & 0 \\ 0 & 0 & 0 \end{bmatrix},$$

$$K_r \frac{d}{dt} \left[K_r^{-1} \right] = \begin{bmatrix} 0 & -(\omega_s - \omega_r) & 0 \\ (\omega_s - \omega_r) & 0 & 0 \\ 0 & 0 & 0 \end{bmatrix},$$

$$K_s L_{ss} K_s^{-1} = \begin{bmatrix} L_s & 0 & 0 \\ 0 & L_s & 0 \\ 0 & 0 & L_{ls} \end{bmatrix},$$

$$K_s L_{sr}' K_r^{-1} = \begin{bmatrix} L_m & 0 & 0 \\ 0 & L_m & 0 \\ 0 & 0 & 0 \end{bmatrix},$$

$$K_r L_{rr}' K_r^{-1} = \begin{bmatrix} L_r & 0 & 0 \\ 0 & L_r & 0 \\ 0 & 0 & L_{ls}' \end{bmatrix},$$

with

$$L_m = \frac{3}{2} L_{ms},$$
$$L_s = L_{ls} + L_m,$$
$$L_r = L_{ls}' + L_m.$$

A.1.2 TORQUE EQUATION IN ARBITRARY REFERENCE-FRAME VARIABLES

The torque equation in the three phase variables of the DFIG is defined as

$$\tau_e = \left(\frac{P}{2} \right) (i_{abcs})^T \frac{\partial}{\partial \theta_r} \left[L_{sr}' \right] i_{abcr}', \tag{A.27}$$

where P is a pair of poles. The term $\frac{\partial}{\partial \theta_r}\left[L'_{sr}\right]$ in (A.27) is defined as

$$\frac{\partial}{\partial \theta_r}\left[L'_{sr}\right] = \begin{bmatrix} -L_{ms}\sin\theta_r & -L_{ms}\sin(\theta_r + \frac{2\pi}{3}) & -L_{ms}\sin(\theta_r - \frac{2\pi}{3}) \\ -L_{ms}\sin(\theta_r - \frac{2\pi}{3}) & -L_{ms}\sin\theta_r & -L_{ms}\sin(\theta_r + \frac{2\pi}{3}) \\ -L_{ms}\sin(\theta_r + \frac{2\pi}{3}) & -L_{ms}\sin(\theta_r - \frac{2\pi}{3}) & -L_{ms}\sin\theta_r \end{bmatrix}.$$

(A.28)

The expression for the electromagnetic torque in terms of arbitrary reference-frame variables may be obtained by substituting the equations of transformation into (A.27). Thus

$$\tau_e = \left(\frac{P}{2}\right)\left(K_s^{-1}i_{dq0s}\right)^T \frac{\partial}{\partial \theta_r}\left[L'_{sr}\right]\left(K_r^{-1}i'_{dq0r}\right).$$

(A.29)

This expression yields the torque expressed in terms of currents as

$$\tau_e = \left(\frac{3}{2}\right)\left(\frac{P}{2}\right)L_m\left(i_{qs}i'_{dr} - i_{ds}i'_{qr}\right).$$

(A.30)

The torque and rotor speed in generator mode are related by

$$\frac{d\omega_r}{dt} = \dot{\omega}_r = \left(\frac{P}{2J}\right)(\tau_m - \tau_e),$$

(A.31)

where τ_m is the mechanical drive torque that the turbine applies to the DFIG, and J is the inertia coefficient.

A.1.3 PER-UNIT CONVERSION

The machine and power system parameters are almost always given in ohms, or percent, or per unit of a base impedance. It is convenient to express the voltage and flux linkage equations in terms of reactances rather than inductances. Hence, (A.26) is often written as

$$v_{dq0s} = -K_s R_s K_s^{-1} i_{dq0s} + \frac{1}{\omega_b} K_s \frac{d}{dt}\left[K_s^{-1}\right] \psi_{dq0s} + \frac{d}{dt}\left[\psi_{dq0s}\right],$$

$$v'_{dq0r} = K_r R'_r K_r^{-1} i'_{dq0r} + \frac{1}{\omega_b} K_r \frac{d}{dt}\left[K_r^{-1}\right] \psi'_{dq0r} + \frac{d}{dt}\left[\psi'_{dq0r}\right], \qquad (A.32)$$

$$\psi_{dq0s} = -\frac{1}{\omega_b} K_s L_{ss} K_s^{-1} i_{dq0s} + \frac{1}{\omega_b} K_s L'_{sr} K_r^{-1} i'_{dq0r},$$

$$\psi'_{dq0r} = -\frac{1}{\omega_b} K_r L'^T_{sr} K_s^{-1} i_{dq0s} + \frac{1}{\omega_b} K_r L'_{rr} K_r^{-1} i'_{dq0r},$$

where

$$\frac{1}{\omega_b} K_s \frac{d}{dt}\left[K_s^{-1}\right] = \begin{bmatrix} 0 & -\frac{\omega_s}{\omega_b} & 0 \\ \frac{\omega_s}{\omega_b} & 0 & 0 \\ 0 & 0 & 0 \end{bmatrix},$$

$$\frac{1}{\omega_b} K_r \frac{d}{dt}\left[K_r^{-1}\right] = \begin{bmatrix} 0 & -\frac{(\omega_s-\omega_r)}{\omega_b} & 0 \\ \frac{(\omega_s-\omega_r)}{\omega_b} & 0 & 0 \\ 0 & 0 & 0 \end{bmatrix},$$

$$\frac{1}{\omega_b} K_s L_{ss} K_s^{-1} = \begin{bmatrix} X_s & 0 & 0 \\ 0 & X_s & 0 \\ 0 & 0 & \frac{L_{ls}}{\omega_b} \end{bmatrix},$$

$$\frac{1}{\omega_b} K_s L'_{sr} K_r^{-1} = \begin{bmatrix} X_m & 0 & 0 \\ 0 & X_m & 0 \\ 0 & 0 & 0 \end{bmatrix},$$

$$\frac{1}{\omega_b} K_r L'_{rr} K_r^{-1} = \begin{bmatrix} X_r & 0 & 0 \\ 0 & X_r & 0 \\ 0 & 0 & \frac{L'_{ls}}{\omega_b} \end{bmatrix},$$

with ω_b as the base electrical angular velocity used to calculate the inductive reactances, as follows:

$$\psi_x = \frac{\lambda_x}{\omega_b},$$

$$X_m = \frac{L_m}{\omega_b},$$

$$X_s = \frac{L_s}{\omega_b},$$

$$X_r = \frac{L_r}{\omega_b}.$$

It is often convenient to express the machine parameters and variables as per-unit quantities. Base power (P_b) and base voltage (V_b) are selected, and all parameters and variables are normalized using these base quantities. The base power may be expressed as

$$P_b = \frac{3}{2} V_b I_b. \tag{A.33}$$

Therefore, because base voltage (V_b) and base power (P_b) are selected, the base current can be calculated from (A.33). It follows that the base impedance may be expressed as

$$Z_b = \frac{V_b}{I_b} = \frac{3V_b^2}{2P_b}. \tag{A.34}$$

The $dq0$ equations written in terms of reactances, (A.32), can be readily converted to per-unit quantities by dividing the voltages by V_b, the currents by I_b, and the resistances and reactances by Z_b.

Although the voltage and flux linkage per-second equations do not change when they are per unitized, the torque equation is modified by the per-unitizing process. For this purpose the base torque may be expressed as

$$\tau_b = \frac{P_b}{(2/P)\omega_b}. \tag{A.35}$$

System (A.32) is rewritten per unit as follows:

$$v_{dq0s(pu)} = -K_s R_s K_{s(pu)}^{-1} i_{dq0s(pu)} + K_s \frac{d}{dt} \left[K_s^{-1}\right]_{(pu)} \Psi_{dq0s(pu)} + \frac{1}{\omega_b} \frac{d}{dt} \left[\Psi_{dq0s(pu)}\right],$$

$$v'_{dq0r(pu)} = K_r R'_r K_{r(pu)}^{-1} i'_{dq0r(pu)} + K_r \frac{d}{dt} \left[K_r^{-1}\right]_{(pu)} \Psi'_{dq0r(pu)} + \frac{1}{\omega_b} \frac{d}{dt} \left[\Psi'_{dq0r(pu)}\right],$$

$$\Psi_{dq0s(pu)} = -K_s L_{ss} K_{s(pu)}^{-1} i_{dq0s(pu)} + K_s L'_{sr} K_{r(pu)}^{-1} i'_{dq0r(pu)},$$

$$\Psi'_{dq0r(pu)} = -K_r L'^T_{sr} K_{s(pu)}^{-1} i_{dq0s(pu)} + K_r L'_{rr} K_{r(pu)}^{-1} i'_{dq0r(pu)}, \qquad \text{(A.36)}$$

where

$$K_s \frac{d}{dt} \left[K_s^{-1}\right]_{(pu)} = \begin{bmatrix} 0 & -1 & 0 \\ 1 & 0 & 0 \\ 0 & 0 & 0 \end{bmatrix},$$

$$K_r \frac{d}{dt} \left[K_r^{-1}\right]_{(pu)} = \begin{bmatrix} 0 & -(1-\omega_{r(pu)}) & 0 \\ (1-\omega_{r(pu)}) & 0 & 0 \\ 0 & 0 & 0 \end{bmatrix},$$

$$K_s L_{ss} K_{s(pu)}^{-1} = \begin{bmatrix} X_{s(pu)} & 0 & 0 \\ 0 & X_{s(pu)} & 0 \\ 0 & 0 & \frac{L_{ls}}{Z_b \omega_b} \end{bmatrix},$$

$$K_s L'_{sr} K_{r(pu)}^{-1} = \begin{bmatrix} X_{m(pu)} & 0 & 0 \\ 0 & X_{m(pu)} & 0 \\ 0 & 0 & 0 \end{bmatrix},$$

$$K_r L'_{rr} K_{r(pu)}^{-1} = \begin{bmatrix} X_{r(pu)} & 0 & 0 \\ 0 & X_{r(pu)} & 0 \\ 0 & 0 & \frac{L'_{ls}}{Z_b \omega_b} \end{bmatrix},$$

where $\omega_b = \omega_s$.

The electric torque τ_e (A.30) and the rotor speed equation (A.31) are rewritten in *pu* as follows:

$$\tau_{e(pu)} = X_{m(pu)} \left(i_{qs(pu)} i'_{dr(pu)} - i_{ds(pu)} i'_{qr(pu)}\right). \qquad \text{(A.37)}$$

$$\frac{d\omega_{r(pu)}}{dt} = \dot{\omega}_{r(pu)} = \left(\frac{1}{2H}\right)\left(\tau_{m(pu)} - \tau_{e(pu)}\right).\tag{A.38}$$

Comment A.1 To facilitate the calculations, we will omit writing the subscript (pu) in all variables used in this book hereafter; the read can take for granted that they are in (pu) unless otherwise stated.

A.1.4 DFIG STATE VARIABLES MODEL

This is essentially an induction machine with wound rotor and variable frequency excitation by the rotor circuit, which is controlled by means of power converters. DFIG configuration allows the rotor speed to vary while synchronizing the stator directly to a fixed frequency power system, the control input is by the rotor winding, and in practice it is possible to measure all the DFIG currents. So, it is convenient to select the stator and rotor currents as DFIG state variables.

Then, in (A.36), substituting ψ_{dq0s} and ψ'_{dq0r} in the v_{dq0s} and v'_{dq0r} equations, respectively,

$$v_{dq0s} = -K_s R_s K_s^{-1} i_{dq0s} + K_s \frac{d}{dt}\left[K_s^{-1}\right]\left(-K_s L_{ss} K_s^{-1} i_{dq0s} + K_s L'_{sr} K_r^{-1} i'_{dq0r}\right)$$
$$+ \frac{d}{dt}\left[-K_s L_{ss} K_s^{-1} i_{dq0s} + K_s L'_{sr} K_r^{-1} i'_{dq0r}\right],\tag{A.39}$$

$$v'_{dq0r} = K_r R'_r K_r^{-1} i'_{dq0r} + K_r \frac{d}{dt}\left[K_r^{-1}\right]\left(-K_r L'^T_{sr} K_s^{-1} i_{dq0s} + K_r L'_{rr} K_r^{-1} i'_{dq0r}\right)$$
$$+ \frac{d}{dt}\left[-K_r L'^T_{sr} K_s^{-1} i_{dq0s} + K_r L'_{rr} K_r^{-1} i'_{dq0r}\right].\tag{A.40}$$

Equations (A.39) and (A.40) can be rewritten as

$$v_{dq0s} = -\left(K_s R_s K_s^{-1} + K_s \frac{d}{dt}\left[K_s^{-1}\right] K_s L_{ss} K_s^{-1}\right) i_{dq0s} + K_s \frac{d}{dt}\left[K_s^{-1}\right] K_s L'_{sr} K_r^{-1} i'_{dq0r}$$
$$- K_s L_{ss} K_s^{-1} \frac{d}{dt}\left[i_{dq0s}\right] + K_s L'_{sr} K_r^{-1} \frac{d}{dt}\left[i'_{dq0r}\right],\tag{A.41}$$

$$v'_{dq0r} = -K_r \frac{d}{dt}\left[K_r^{-1}\right] K_r L_{sr}^{'T} K_s^{-1} i_{dqos} + \left(K_r R'_r K_r^{-1} + K_r \frac{d}{dt}\left[K_r^{-1}\right] K_r L'_{rr} K_r^{-1}\right) i'_{dq0r}$$

$$-K_r L_{sr}^{'T} K_s^{-1} \frac{d}{dt}\left[i_{dqos}\right] + K_r L'_{rr} K_r^{-1} \frac{d}{dt}\left[i'_{dqor}\right]. \tag{A.42}$$

In order to simplify the handling, Equations (A.41) and (A.42) can be rewritten in matrix form as

$$\begin{bmatrix} v_{dq0s} \\ v'_{dq0r} \end{bmatrix} = \begin{bmatrix} -K_s R_s K_s^{-1} - K_s \frac{d}{dt}\left[K_s^{-1}\right] K_s L_{ss} K_s^{-1} & K_s \frac{d}{dt}\left[K_s^{-1}\right] K_s L'_{sr} K_r^{-1} \\ -K_r \frac{d}{dt}\left[K_r^{-1}\right] K_r L_{sr}^{'T} K_s^{-1} & K_r R'_r K_r^{-1} + K_r \frac{d}{dt}\left[K_r^{-1}\right] K_r L'_{rr} K_r^{-1} \end{bmatrix}$$

$$\begin{bmatrix} i_{dq0s} \\ i'_{dq0r} \end{bmatrix} + \begin{bmatrix} -K_s L_{ss} K_s^{-1} & K_s L'_{sr} K_r^{-1} \\ -K_r L_{sr}^{'T} K_s^{-1} & K_r L'_{rr} K_r^{-1} \end{bmatrix} \frac{d}{dt} \begin{bmatrix} i_{dq0s} \\ i'_{dq0r} \end{bmatrix}, \tag{A.43}$$

where each matrix term is defined as

$$-K_s R_s K_s^{-1} - K_s \frac{d}{dt}\left[K_s^{-1}\right] K_s L_{ss} K_s^{-1} = \begin{bmatrix} -r_s & X_s & 0 \\ -X_s & -r_s & 0 \\ 0 & 0 & -r_s \end{bmatrix}, \tag{A.44}$$

$$K_s \frac{d}{dt}\left[K_s^{-1}\right] K_s L'_{sr} K_r^{-1} = \begin{bmatrix} 0 & -X_m & 0 \\ X_m & 0 & 0 \\ 0 & 0 & 0 \end{bmatrix}, \tag{A.45}$$

$$-K_r \frac{d}{dt}\left[K_r^{-1}\right] K_r L_{sr}^{'T} K_s^{-1} = \begin{bmatrix} 0 & -X_m(\omega_r - 1) & 0 \\ X_m(\omega_r - 1) & 0 & 0 \\ 0 & 0 & 0 \end{bmatrix}, \tag{A.46}$$

$$K_r R'_r K_r^{-1} + K_r \frac{d}{dt}\left[K_r^{-1}\right] K_r L'_{rr} K_r^{-1} = \begin{bmatrix} r'_r & X_r(\omega_r - 1) & 0 \\ -X_r(\omega_r - 1) & r'_r & 0 \\ 0 & 0 & r'_r \end{bmatrix}. \tag{A.47}$$

Then Equation (A.43) can be rewritten as

$$v_{dq0} = Z i_{dq0} + L \frac{d}{dt} i_{dq0}, \tag{A.48}$$

where

$$v_{dq0} = \begin{bmatrix} v_{ds} \\ v_{qs} \\ v_{0s} \\ v'_{dr} \\ v'_{qr} \\ v'_{0r} \end{bmatrix}, i_{dq0} = \begin{bmatrix} i_{ds} \\ i_{qs} \\ i_{0s} \\ i'_{dr} \\ i'_{qr} \\ i'_{0r} \end{bmatrix}, \tag{A.49}$$

$$Z = \begin{bmatrix} -r_s & X_s & 0 & 0 & -X_m & 0 \\ -X_s & -r_s & 0 & X_m & 0 & 0 \\ 0 & 0 & -r_s & 0 & 0 & 0 \\ 0 & -X_m(\omega_r - 1) & 0 & r'_r & X_r(\omega_r - 1) & 0 \\ X_m(\omega_r - 1) & 0 & 0 & -X_r(\omega_r - 1) & r'_r & 0 \\ 0 & 0 & 0 & 0 & 0 & r'_r \end{bmatrix}, \tag{A.50}$$

$$L = \begin{bmatrix} -\frac{X_s}{\omega_b} & 0 & 0 & \frac{X_m}{\omega_b} & 0 & 0 \\ 0 & -\frac{X_s}{\omega_b} & 0 & 0 & \frac{X_m}{\omega_b} & 0 \\ 0 & 0 & -\frac{L_{ls}}{\omega_b} & 0 & 0 & 0 \\ -\frac{X_m}{\omega_b} & 0 & 0 & \frac{X_r}{\omega_b} & 0 & 0 \\ 0 & -\frac{X_m}{\omega_b} & 0 & 0 & \frac{X_r}{\omega_b} & 0 \\ 0 & 0 & 0 & 0 & 0 & \frac{L'_{ls}}{\omega_b} \end{bmatrix}. \tag{A.51}$$

In (A.48), it is easy to solve $\frac{d}{dt}i_{dq0}$ as follows:

$$\frac{d}{dt}I_{dq0} = -L^{-1}ZI_{dq0} + L^{-1}V_{dq0}. \tag{A.52}$$

The main feature of the transformation to a frame fixed in the rotor $(d - q)$ is that the variables that belong to the 0 axis are independent of ω; then these are not associated with the transformation frame. Additionally, the variables i_{0s}, i_{0r}, v_{0s}, and

v_{0r} are 0 for a balanced system. Then, system (A.52) can be reduced to

$$\frac{d}{dt} i_{dq} = A(\omega_r) i_{dq} + B v_{dq},\qquad\qquad (A.53)$$

where

$$v_{dq} = \begin{bmatrix} v_{ds} \\ v_{qs} \\ v'_{dr} \\ v'_{qr} \end{bmatrix}, i_{dq} = \begin{bmatrix} i_{ds} \\ i_{qs} \\ i'_{dr} \\ i'_{qr} \end{bmatrix},$$

$$A(\omega_r) = \begin{bmatrix} -\frac{\omega_b r_s}{X_s \sigma} & \omega_b\left(1 - \frac{\sigma-1}{\sigma}\omega_r\right) & -\frac{\omega_b X_m r'_r}{X_s X_r \sigma} & -\frac{\omega_b X_m}{X_s \sigma}\omega_r \\ -\omega_b\left(1 - \frac{\sigma-1}{\sigma}\omega_r\right) & -\frac{\omega_b r_s}{X_s \sigma} & \frac{\omega_b X_m}{X_s \sigma}\omega_r & -\frac{\omega_b X_m r'_r}{X_s X_r \sigma} \\ -\frac{\omega_b X_m r_s}{X_s X_r \sigma} & \frac{\omega_b X_m}{X_r \sigma}\omega_r & -\frac{\omega_b r'_r}{X_r \sigma} & \omega_b\left(1 - \frac{1}{\sigma}\omega_r\right) \\ -\frac{\omega_b X_m}{X_r \sigma}\omega_r & -\frac{\omega_b X_m r_s}{X_s X_r \sigma} & -\omega_b\left(1 - \frac{1}{\sigma}\omega_r\right) & -\frac{\omega_b r'_r}{X_r \sigma} \end{bmatrix},$$

$$B = \begin{bmatrix} \frac{-\omega_b}{X_s \sigma} & 0 & \frac{\omega_b X_m}{X_s X_r \sigma} & 0 \\ 0 & \frac{-\omega_b}{X_s \sigma} & 0 & \frac{\omega_b X_m}{X_s X_r \sigma} \\ -\frac{\omega_b X_m}{X_s X_r \sigma} & 0 & \frac{\omega_b}{X_r \sigma} & 0 \\ 0 & -\frac{\omega_b X_m}{X_s X_r \sigma} & 0 & \frac{\omega_b}{X_r \sigma} \end{bmatrix},$$

with

$$\sigma = 1 - \frac{X_m^2}{X_s X_r}.$$

Equations (A.38) and (A.53) are the state space representation of the DFIG. This representation has four electrical variables (i_{ds}, i_{qs}, i_{dr}, i_{qr}) and one mechanical variable (ω_r).

A.2   DC LINK MATHEMATICAL MODEL

The DFIG allows the rotor speed to vary while synchronizing the stator directly to a fixed frequency power system, which is achieved by controlling the rotor side converter (RSC). The RSC is connected via a DC Link to a grid side converter (GSC), which is in turn connected to the stator terminals directly or through a step-up transformer. The circuit of the DC Link connected to the GSC and the last one connected to the stator terminals can be considered as a STATCOM, as shown in Figure A.4. The GSC block in the circuit is treated as an ideal. Assuming balanced conditions,

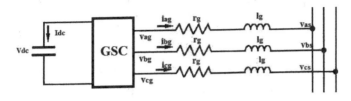

FIGURE A.4   DC Link configuration (STATCOM).

the AC-side circuit equations in Figure A.4 can be written as

$$v_{ag} - v_{as} = r_g i_{ag} + \frac{d\lambda_{ag}}{dt},$$
$$v_{bg} - v_{bs} = r_g i_{bg} + \frac{d\lambda_{bg}}{dt}, \qquad\qquad (A.54)$$
$$v_{cg} - v_{cs} = r_g i_{cg} + \frac{d\lambda_{cg}}{dt},$$

$$\lambda_{ag} = l_g i_{ag},$$
$$\lambda_{bg} = l_g i_{bg}, \qquad\qquad (A.55)$$
$$\lambda_{ag} = l_g i_{ag},$$

where i_{ag}, i_{bg}, and i_{cg} are the currents in the abc axis, respectively; v_{as}, v_{bs}, and v_{cs} are the three-phase grid voltages; and v_{ag}, v_{bg}, and v_{cg} are the three-phase voltages generated by the GSC, which are the control input for the DC Link circuit. Equations (A.54) and (A.55) can be rewritten in matrix form as follows:

$$v_{abdcg} - v_{abcs} = R_g i_{abcg} + \frac{d\lambda_{abcg}}{dt}, \qquad\qquad (A.56)$$

$$\lambda_{abcg} = L_g i_{abcg}, \qquad\qquad (A.57)$$

where

$$v_{abcg} = \begin{bmatrix} v_{ag} \\ v_{bg} \\ v_{cg} \end{bmatrix}, \quad v_{abcs} = \begin{bmatrix} v_{as} \\ v_{bs} \\ v_{cs} \end{bmatrix}, \quad i_{abcg} = \begin{bmatrix} i_{ag} \\ i_{bg} \\ i_{cg} \end{bmatrix}, \quad \lambda_{abcg} = \begin{bmatrix} \lambda_{ag} \\ \lambda_{bg} \\ \lambda_{cg} \end{bmatrix},$$

$$R_g = \begin{bmatrix} r_g & 0 & 0 \\ 0 & r_g & 0 \\ 0 & 0 & r_g \end{bmatrix}, \quad L_g = \begin{bmatrix} l_g & 0 & 0 \\ 0 & l_g & 0 \\ 0 & 0 & l_g \end{bmatrix}.$$

By the same reasons described in Subsection A.1.1, systems (A.2) and (A.56) can be written in a synchronously rotating reference frame using the d-q transformation. Then the used transformation is defined as

$$K_s = \frac{2}{3} \begin{bmatrix} \cos\theta & \cos\left(\theta - \frac{2\pi}{3}\right) & \cos\left(\theta + \frac{2\pi}{3}\right) \\ -\sin\theta & -\sin\left(\theta - \frac{2\pi}{3}\right) & -\sin\left(\theta + \frac{2\pi}{3}\right) \\ \frac{1}{2} & \frac{1}{2} & \frac{1}{2} \end{bmatrix}, \tag{A.58}$$

$$K_s^{-1} = \begin{bmatrix} \cos\theta & -\sin\theta & 1 \\ \cos(\theta - \frac{2\pi}{3}) & -\sin(\theta - \frac{2\pi}{3}) & 1 \\ \cos(\theta + \frac{2\pi}{3}) & -\sin(\theta + \frac{2\pi}{3}) & 1 \end{bmatrix}. \tag{A.59}$$

Now, applying the transformation to Equations (A.56) and (A.2), the follow equation is obtained:

$$K_s^{-1} v_{dq0g} - K_s^{-1} v_{dq0s} = R_g K_s^{-1} i_{dq0g} + \frac{d}{dt}\left(K_s^{-1} \lambda_{dq0g}\right)$$

$$K_s^{-1} \lambda_{dq0g} = L_g K_s^{-1} i_{dq0g}$$

and, finally, the change of variable is derived as

$$v_{dq0g} - v_{dq0s} = K_s R_g K_s^{-1} i_{dq0g} + K_s \frac{d}{dt}\left(K_s^{-1}\right) \lambda_{dq0g} + \frac{d}{dt}\left(\lambda_{dq0g}\right), \tag{A.60}$$

$$\lambda_{dq0g} = K_s L_g K_s^{-1} i_{dq0g}, \tag{A.61}$$

where

$$K_s R_g K_s^{-1} = R_g = \begin{bmatrix} r_g & 0 & 0 \\ 0 & r_g & 0 \\ 0 & 0 & r_g \end{bmatrix},$$

$$K_s \frac{d}{dt} \left[K_s^{-1} \right] = \begin{bmatrix} 0 & -\omega & 0 \\ \omega & 0 & 0 \\ 0 & 0 & 0 \end{bmatrix},$$

$$K_s L_g K_s^{-1} = L_g = \begin{bmatrix} l_g & 0 & 0 \\ 0 & l_g & 0 \\ 0 & 0 & l_g \end{bmatrix}.$$

Substitute (A.61) in (A.60) as follows:

$$v_{dq0g} - v_{dq0s} = R_g i_{dq0g} + K_s \frac{d}{dt} \left(K_s^{-1} \right) L_g i_{dq0g} + \frac{d}{dt} \left(L_g i_{dq0g} \right). \tag{A.62}$$

Now, in (A.62) solving for $\dfrac{di_{dqg}}{dt}$, the following equation is obtained:

$$\frac{di_{dqg}}{dt} = L_g^{-1} \left(-R_g i_{dq0g} - K_s \frac{d}{dt} \left(K_s^{-1} \right) L_g i_{dq0g} + v_{dq0g} - v_{dq0s} \right). \tag{A.63}$$

As in Subsection A.1.1, the transformed variables that belong to the 0 axis are 0 for a balanced system. Then, the equation obtained above can be written as

$$\frac{di_{dqg}}{dt} = A_g i_{dqg} + B_g v_{dqg} - B_g v_{dqs}, \tag{A.64}$$

where

$$A_g = -L_g^{-1} R_g - L_g^{-1} K_s \frac{d}{dt} \left(K_s^{-1} \right) L_g$$

$$= \begin{bmatrix} -\dfrac{r_g}{l_g} & \omega \\ -\omega & -\dfrac{r_g}{l_g} \end{bmatrix}, \tag{A.65}$$

$$B_g = L_g^{-1}$$

$$= \begin{bmatrix} \dfrac{1}{l_g} & 0 \\ 0 & \dfrac{1}{l_g} \end{bmatrix}, \tag{A.66}$$

$$i_{dqg} = \begin{bmatrix} i_{dg} \\ i_{qg} \end{bmatrix}, \quad v_{dqg} = \begin{bmatrix} v_{dg} \\ v_{qg} \end{bmatrix}, \quad v_{dqs} = \begin{bmatrix} v_{ds} \\ v_{qs} \end{bmatrix},$$

where (i_{dg}, i_{qg}), (v_{dg}, v_{qg}), and (v_{ds}, v_{qs}) are the d-q components of (i_{ag}, i_{bg}, i_{cg}), (v_{ag}, v_{bg}, v_{cg}), and (v_{as}, v_{bs}, v_{cs}), respectively.

Neglecting the harmonics due to switching and the losses in the GSC and the transformer, the power balance between the AC and DC sides of the GSC is given by

$$\frac{3}{2} \left(v_{ds} i_{dg} + v_{qs} i_{qg} \right) = v_{dc} i_{dc} = C v_{dc} \frac{dv_{dc}}{dt}, \tag{A.67}$$

where solving for $\dfrac{dv_{dc}}{dt}$ the following equation is obtained:

$$\frac{dv_{dc}}{dt} = \frac{3}{2Cv_{dc}} \left(v_{ds} i_{dg} + v_{qs} i_{qg} \right). \tag{A.68}$$

Equations (A.68) and (A.64) are the state space representation of the DC Link. This representation has three electrical variables (v_{dc}, i_{dg}, i_{qr}).

Based on Section A.1.3, the DC Link mathematical model in pu is obtained as

$$\frac{dv_{dc(pu)}}{dt} = \frac{1}{Cv_{dc(pu)}} v_{dqs(pu)}^T M_{p(pu)} v_{dqg(pu)}, \tag{A.69}$$

$$\frac{di_{dqg(pu)}}{dt} = A_{g(pu)} i_{dqg(pu)} + B_{g(pu)} v_{dqg(pu)} - B_{g(pu)} v_{dqs(pu)}, \tag{A.70}$$

where

$$M_{p(pu)} = \begin{bmatrix} 1 & 0 \\ 0 & 1 \end{bmatrix}, \quad A_{g(pu)} = \begin{bmatrix} -\dfrac{\omega_b r_g}{X_l} & \omega_s \\ -\omega_s & -\dfrac{\omega_b r_g}{X_l} \end{bmatrix}, \quad B_{g(pu)} = \begin{bmatrix} \dfrac{\omega_b}{X_l} & 0 \\ 0 & \dfrac{\omega_b}{X_l} \end{bmatrix}.$$

In Section 6.1, the neural identifier structures are described, and the DFIG and DC Link neural identifiers are included in Subsection 6.1.1 and 6.1.2.

In Section 6.2, the sliding modes scheme based on the neural identifiers is presented. In Subsection 6.2.1, the neural block control scheme is used to design the DFIG controller. Additionally, simulation results are presented to validate the control performance. In Subsection 6.2.2, a similar neural network scheme is applied to the DC Link and the corresponding simulation results are presented.

In Section 6.3, the inverse optimal control scheme based on the neural mathematical model is presented. In Subsection 6.3.1, the DFIG neural inverse optimal controller is developed, and Subsection 6.3.2 presents the DC Link neural inverse optimal controller. Both of them include the corresponding simulation results.

Index